Converging Minds

This groundbreaking book explores the power of collaborative AI in amplifying human creativity and expertise. Written by two seasoned experts in data analytics, AI, and machine learning, the book offers a comprehensive overview of the creative process behind AI-powered content generation. It takes the reader through a unique collaborative process between human authors and various AI-based topic experts, created, prompted, and fine-tuned by the authors.

This book features a comprehensive list of prompts that readers can use to create their own ChatGPT-powered topic experts. By following these expertly crafted prompts, individuals and businesses alike can harness the power of AI, tailoring it to their specific needs and fostering a fruitful collaboration between humans and machines. With real-world use cases and deep insights into the foundations of generative AI, the book showcases how humans and machines can work together to achieve better business outcomes and tackle complex challenges. Social and ethical implications of collaborative AI are covered and how it may impact the future of work and employment. Through reading the book, readers will gain a deep understanding of the latest advancements in AI and how they can shape our world.

Converging Minds: The Creative Potential of Collaborative AI is essential reading for anyone interested in the transformative potential of AI-powered content generation and human-AI collaboration. It will appeal to data scientists, machine learning architects, prompt engineers, general computer scientists, and engineers in the fields of generative AI and deep learning.

Aleksandra Przegalinska is an Associate Professor and Vice-President of Kozminski University, responsible for international relations and Ethics and Social Responsibility (ESR) as well as Senior Research Associate at the Harvard Labour and Worklife Program. Aleksandra is the Head of the Human-Machine Interaction Research Center at Kozminski University and the Leader of the AI in Management Program. Until recently, she conducted postdoctoral research at the Center for Collective Intelligence at the Massachusetts Institute of Technology in Boston. She graduated from The New School for Social Research in New York. She is the co-author of *Collaborative Society* (The MIT Press) and *Strategizing AI in Business and Education* (Cambridge University Press) published together with Dariusz Jemielniak.

Tamilla Triantoro is an Associate Professor in the School of Business at Quinnipiac University. She has directed graduate and undergraduate programs in Business Analytics at Quinnipiac University and the University of Connecticut. Her expertise includes artificial intelligence, human-AI collaboration, and the future of work. She has spoken about these topics in various parts of the world and presented her work on six continents. With a PhD from the City University of New York, where she researched online user behavior, Dr. Triantoro brings a deep understanding of the human element to her work.

Human Factors in Design, Engineering, and Computing

Series Editors: Tareq Ahram and Waldemar Karwowski

This series focuses on research and development efforts intended to promote the comprehensive integration of people and technological systems. To help foster integration of human factors research with emerging technologies, this series will focus on novel methodologies, design tools, and solutions that advance our understanding of the nature of human factors and ergonomics with intelligent technologies and services. It will address all aspects of human factors and human-centered design, with a particular emphasis on applications of emerging technologies, computing, artificial intelligence, and systems. The series offers a multidisciplinary platform for researchers and practitioners alike, discussing emerging issues in the field of human factors engineering systems, with a special focus on (but not limited to) computing and AI-based technologies. It will be useful for researchers, senior graduate students, graduate students, and professionals in different domains including human factors, engineering design, systems science and engineering, and ergonomics.

Forthcoming titles

Human Factors in Design: Intelligent Tools and Technological Innovations
Edited by Tareq Ahram and Waldemar Karwowski

Human Factors in Engineering: Manufacturing Systems, Automation, and Interactions
Edited by Beata Mrugalska, Tareq Ahram and Waldemar Karwowski

Human Factors in Computing: Social Dimension and Artificial Intelligence-Based Technologies
Edited by Tareq Ahram and Waldemar Karwowski

Human Factors and Emerging Technologies: Integrating People and Technological Systems
Edited by Tareq Ahram and Waldemar Karwowski

Human Factors in Intelligence, Technology and Analytics
Edited by Tareq Ahram and Waldemar Karwowski

Human-Technology: Blending Artificial Intelligence, Computing, and Intelligent Design
Edited by Tareq Ahram and Waldemar Karwowski

Converging Minds: The Creative Potential of Collaborative AI
Aleksandra Przegalinska and Tamilla Triantoro

For more information about this series, please visit: https://www.routledge.com/Human-Factors-In-Design-Engnineering-and-Computing/book-series/%20CRCHFIDEC

Converging Minds
The Creative Potential
of Collaborative AI

Aleksandra Przegalinska and Tamilla Triantoro

CRC Press
Taylor & Francis Group
Boca Raton London New York

CRC Press is an imprint of the
Taylor & Francis Group, an **informa** business

Designed cover image: © Shutterstock, illustration contributor: agsandrew

First edition published 2024
by CRC Press
2385 NW Executive Center Drive, Suite 320, Boca Raton FL 33431

and by CRC Press
4 Park Square, Milton Park, Abingdon, Oxon, OX14 4RN

CRC Press is an imprint of Taylor & Francis Group, LLC

ISBN: 978-1-032-62687-1 (hbk)
ISBN: 978-1-032-65660-1 (pbk)
ISBN: 978-1-032-65661-8 (ebk)

DOI: 10.1201/9781032656618

Typeset in Times
by codeMantra

Contents

Acknowledgments

The journey of writing this book has been a collaborative endeavor, enriched by the insights, expertise, and dedication of remarkable individuals. With a sense of profound gratitude, we acknowledge the contributions of those who have been instrumental in bringing this work to fruition.

Gedeon Werner has been the cohesive force of this project. His ability to unify diverse minds towards a common goal has been nothing short of exemplary.

James Hobbs, our diligent publishing agent, has been a pillar of support, always available and ensuring a smooth transition from manuscript to published work.

Balaji Padmanabhan generously shared his illuminating thoughts, adding a layer of depth to our discourse.

Leon Ciechanowski provided a solid foundation invaluable in navigating the technical landscapes explored in the book.

Dariusz Jemielniak provided fresh perspectives that enhanced the quality of our deliberations.

Guido Lang demonstrated meticulous attention to detail. His insights have been invaluable to the quality of this work.

Aureliusz Gorski has been a living testament to the transformative power of human-AI collaboration. His journey and insights have been pivotal in shaping the narrative of this book.

Ardian Triantoro has been a catalyst in propelling the thought process forward, consistently bringing innovative ideas to the table.

Alicja Skierkowska, Aleksandra's daughter, beacon and muse, is and has been the unseen wavelength in the spectrum of Aleksandra's work.

The convergence of these diverse and insightful minds has made the journey of creating this book an enlightening and rewarding experience. Our sincere appreciation extends to each one of them for their indispensable contributions.

Aleksandra Przegalinska's work on this book was supported by the National Science Center (NCN) grant 516876.

Prologue

In the midst of the global pandemic, we were brought together by our shared passion for exploring human-AI collaboration. Our initial online interactions were mediated by computer screens. As the world slowly began to open, we seized the opportunity to meet in person in a quaint cafe near Harvard University in Cambridge, Massachusetts. Over a cup of coffee we shared our research and aspirations, and that encounter was the beginning of our collaborative exploration of AI.

Our workplaces are Kozminski University and Quinnipiac University which are separated by the ocean: one is located in Europe, and another one is in the United States. However, the distance and time difference did not stop us from collaboration. Our first joint work was our blog, "AI ONE ON ONE." With each iteration of Generative Pre-Trained Transformer (GPT), from GPT-3 to GPT-3.5 and then to 4.0, and new offerings such as Claude and Gemini, we explored, challenged, and expanded our understanding of the potential of generative AI. Through our shared journey, we have come to appreciate that true collaboration, whether between humans or with machines, knows no boundaries.

As artificial intelligence continues to transform our lives, it has become increasingly clear that effective human-machine collaboration will be critical to unlocking our full potential. This book aims to provide a comprehensive overview of converging minds, that comprises collaborative AI, its applications, and its implications for the future. The book is intended to be a resource for students, researchers, and professionals who are interested in understanding the latest developments in collaborative AI and how it can be used to drive innovation and address complex challenges. Drawing on our expertise in data analytics, human-AI collaboration, and machine learning, we have designed the book to be accessible to a wide range of readers, regardless of their technical background.

Throughout the book, we explore the foundations of collaborative AI, including the generative AI technologies such as Generative Adversarial Networks (GANs) and Variational Autoencoders (VAEs). We discuss the practical applications of collaborative AI in creative work, education, scientific research, and business, as well as its implications for the future of work and the labor market. We offer ideas of augmenting human abilities, discuss strategies of human unicorns, and attempt predicting the future using AI. Additionally, we examine the social and ethical implications of collaborative AI, including Artificial General Intelligence (AGI), their potential and risks.

We sincerely hope that this book will serve as a valuable resource for anyone interested in understanding the latest developments in collaborative AI and how it can be used to drive innovation and address complex challenges in a more effective and sustainable way.

Aleksandra Przegalinska   Tamilla Triantoro

Introduction

The Promise and Peril of Collaborative AI

In the rapidly evolving area of artificial intelligence, we, as co-authors, wish to make a clear stance at the outset: while AI as an assistive technology can be very beneficial for humanity, the pursuit of artificial general intelligence (AGI) (Goertzel, 2016; Loos, 2023) holds less allure for us. By AGI we understand the type of artificial intelligence that possesses the ability to understand, learn, and apply knowledge across different domains, reason through problems, have consciousness, and even have emotional understanding. Unlike narrow or specialized AI, which is designed for specific tasks, AGI would have the general problem-solving capabilities comparable to human intelligence (Kurzweil, n.d.; Bostrom, 2014; Hutter, 2005). Here, however, one should add that the very definition of AGI is currently challenged by various technological companies, like OpenAI, that are more likely to define AGI as a system that surpasses human capabilities in the majority of economically valuable tasks (OpenAI, 2023). We make the case that making a distinction between current generative tools and AGI defined in the latter way is rather impossible. Already existing generative tools are or will be soon on this level and it is certainly not comparable with AGI as it was defined prior.

Let us come back to the "historical" definition of AGI and ruminate where it could take us. While the idea of machines mirroring the full spectrum of human intelligence is tantalizing for many, we find at least a couple of compelling reasons to diverge from this path. First, the very nature of AGI, which seeks to replicate humanlike cognition in its entirety, raises profound ethical, philosophical, and practical challenges. The unpredictability of such a system, coupled with the inherent risks of unintended consequences, makes its pursuit a precarious endeavor. Second, the vast resources dedicated to achieving AGI could be better utilized in addressing more immediate and tangible challenges that narrow AI can help solve (Sabry, 2023).

Instead, our fascination lies with artificial intelligence as a tool, an extension of the human mind. We see immense value in AI that, while narrow in its scope, is adept and precise in its function. Such AI, when fed with quality data, becomes a powerful ally, enhancing our capabilities, offering new perspectives, and driving innovation. It serves as a sparring partner, challenging our assumptions and pushing the boundaries of what we know. It collaborates with us, not as a replacement but as a partner, bringing its unique strengths to the table. We believe that ChatGPT and other large language models can, if properly directed, serve exactly this function and become our helpers and cognitive extensions (Lukowicz et al., 2023).

DOI: 10.1201/9781032656618-1

In this book, we aim to go deep into the potential of the collaborative AI (Koch & Oulasvirta, 2018). We argue that the true promise of AI lies not in mimicking human intelligence in its entirety but in complementing it. The synergy between human creativity and the computational prowess of still narrow, but smart AI, is where the future beckons. While the allure of AGI will undoubtedly continue to captivate many, we believe that the real transformative potential lies in the here and now, with the AI technologies that are already reshaping our world.

The quest for AGI (Goertzel et al., 2022; Schmidhuber et al., 2011), though ambitious, sometimes overshadows the profound advancements and possibilities that are already within our grasp. It is in this more accessible sphere of AI, where machines don't replace but rather augment human capabilities, that we find the most promising and tangible benefits (Ludik, 2021). This brings us to the concept of collaborative AI, a paradigm that encapsulates the essence of human-machine synergy. Instead of striving for an elusive and complete replication of human intelligence, collaborative AI emphasizes the power of partnership, where both human and machine agents contribute their unique strengths to achieve shared objectives. In general, collaborative AI that we refer to so often in this book is a paradigm of artificial intelligence that emphasizes the synergistic interaction between human and machine agents. Rather than viewing AI as an isolated system or a tool, collaborative AI positions it as an active participant in a joint endeavor with humans. This collaboration is characterized by mutual influence, shared goals, and dynamic feedback loops (Feldman, 2017; Brynjolfsson & McAfee, 2014).

In the manner following the collaborative AI paradigm, this book didn't generate itself. Instead, it is a result of human-AI collaboration. It follows the paradigm of converging minds with a hope to contribute to the further development of this paradigm. At the heart of collaborative, AI is the recognition of the complementary nature of human and machine intelligence. While humans excel in creativity, intuition, and contextual understanding, AI brings computational efficiency, pattern recognition, and data-driven insights (Borowik et al., 2015). The collaboration between the two is not additive but multiplicative, leading to outcomes that neither could achieve independently. Furthermore, collaborative AI acknowledges the importance of trust, transparency, and ethical considerations in human-machine interactions. It seeks to create systems that are not only technically proficient but also align with human values, norms, and expectations.

We hope that our book introduces a distinctive approach by elucidating the various modes of collaborating with AI (Sowa et al., 2021): parallel work, synergy, contributive work, team work, and hybrid models, to name just a few. While many discussions on AI focus on its technical prowess or its potential societal impacts, few consider the nuanced ways in which humans can engage with AI to foster a more productive and harmonious partnership. This exploration is not only theoretical; it is the very ethos of our book.

What sets our work apart is the articulation of these collaborative modes and the lived experience of them. As we start the journey of crafting this book, we are not just authors elucidating concepts; we are also striving practitioners that embody these modes. Each chapter, each section, and indeed each page is the evidence of the dynamic interplay between human intellect and AI capabilities. Whether it's

AI amplifying our stylistic choices, challenging our assertions, or aiding in the generation of novel ideas, readers will witness firsthand the transformative potential of these collaborative modes.

By intertwining theory with practice, our book (hopefully) offers readers a dual experience: a deeper intellectual understanding of human-AI collaboration and a tangible demonstration of its potential. In doing so, we aim to not only inform but also inspire, showcasing how the convergence of human and machine intelligence can lead to unparalleled creative and scholarly achievements.

THE IMPERATIVE OF COLLABORATION

Throughout human history, few things have propelled us forward more than our ability to collaborate. Collective human effort has been the cornerstone of our most awe-inspiring achievements. The journey from cave wall scribblings to today's sophisticated large language models demonstrates human unquenchable thirst for progress. Our early ancestors used rudimentary symbols to represent tangible objects or basic ideas, essentially laying the groundwork for written language. Over time, this led to the development of alphabets and grammatical rules, which were eventually codified in books, first handwritten, then printed, thanks to inventions like the Gutenberg press (McLuhan et al., 2011; Naughton, 2014). The digital age turbocharged this progression, enabling the creation of vast databases and the development of algorithms to parse and analyze them.

Throughout these periods, the driving force has been a uniquely human blend of curiosity and collaboration. We accumulated knowledge and built systems to categorize, share, and critique it, refining our collective understanding over generations. Today, large language models showcase this endeavor, complex algorithms trained on the sum total of human text published online, capable of generating coherent and contextually relevant language (Chang et al., 2023). These models are a milestone in humanity's ongoing quest to understand and interact with the world through the power of language.

Collaboration is a complex and multifaceted phenomenon in the human-human context (Jemielniak & Przegalinska, 2019), encompassing a wide range of interactions, relationships, and processes. It is not only a matter of working together but also a deeper level of engagement, communication, trust, and shared understanding.

In the context of human collaboration, many authors (Koschmann, 2012; Stahl, 2010) emphasize the importance of shared meaning-making, where collaborators actively construct and negotiate meaning through interaction. In Chapter 3, the readers will find a lot of content on collaboration with AI and the future of work, and in Chapter 5, we will talk about our hypothetical future with generative AI and beyond it. This process is not just about exchanging information but involves a dynamic interplay of perspectives, interpretations, and intentions. Trust is another critical element of human collaboration (Shneiderman, 2022; Kumar & Muneesh, 2011). Trust enables individuals to take risks, share ideas, and open themselves to vulnerability, knowing that their collaborators will act in a supportive and responsible manner. Building and maintaining trust requires ongoing effort, communication, and empathy.

Communication, as highlighted by Fussell and Kreuz (2014), is not just a tool for transmitting information but a complex process of encoding, decoding, interpreting, and understanding. Effective communication in collaboration requires awareness of verbal and nonverbal cues, cultural norms, individual preferences, and the broader context in which the collaboration takes place. Moreover, the role of leadership in collaboration is also significant. Leadership in a collaborative context is not about command and control but facilitating, guiding, and supporting the collaborative process (Baxter & Koehler, 2013). It involves creating an environment where collaboration can thrive, recognizing individual strengths, fostering diversity, and aligning goals. It is particularly visible in the emerging governance models of peer production and open collaboration (Jemielniak, 2014).

Collaboration is also shaped by the sociocultural context, as argued by Vygotsky (1980). Social norms, cultural values, historical background, and institutional structures all influence how collaboration is understood, practiced, and valued. Collaboration is not a universal or static concept but situated and evolves over time.

Furthermore, collaboration is not without challenges and tensions. As identified by Hargreaves and Fullan (2015), collaboration can lead to conflicts, power dynamics, misunderstandings, and ethical dilemmas. Navigating these challenges requires emotional intelligence, reflexivity, negotiation skills, and ethical awareness.

Collaboration in the human-human context is a rich and complex phenomenon that goes beyond cooperation or coordination. It involves shared meaning-making, trust, communication, leadership, socio-cultural context, and ethical considerations. Understanding collaboration in its depth and complexity is essential for both theoretical exploration and practical application in various fields, from education and business to healthcare and community development (Mahajan, 2021; Anderson & Coveyduc, 2020).

HUMAN-MACHINE COLLABORATION

Human-machine collaboration represents a significant shift in the way humans interact with technology, moving beyond tool usage to a more integrated and interactive relationship. This collaboration is not a recent phenomenon, but has historical roots that reflect the evolving capabilities of machines and the changing understanding of human cognition and agency.

Historically, machines were seen primarily as tools that extended human physical capabilities. The industrial revolution marked a significant milestone in this regard, with machines taking over manual labor and increasing efficiency (Brynjolfsson & McAfee, 2014). However, the relationship was largely unidirectional, with humans controlling machines and machines passively executing tasks. With the advent of computers and artificial intelligence, the nature of human-machine collaboration began to change. Machines started to exhibit cognitive-like abilities, such as problem-solving, learning, and adaptation (Russell & Norvig, 2016). This shift led to a more interactive and reciprocal relationship, where machines could respond to human input, provide feedback, and even influence human decision-making.

The development of collaborative robots or "cobots" further advanced human-machine collaboration, allowing for physical interaction and cooperation in shared spaces (Bicchi et al., 2008). Cobots are designed to work alongside humans, recognize and adapt to human actions, and foster a partnership rather than task execution. We feel that the concept of "symbiotic autonomy," as articulated by Coradeschi (2013), captures the essence of contemporary human-machine collaboration. It emphasizes mutual dependence, where humans and machines rely on each other's strengths and compensate for each other's weaknesses. This symbiosis requires sophisticated machine understanding of human intentions, emotions, and context, as well as ethical considerations regarding autonomy, responsibility, and trust.

The ethical dimension of human-machine collaboration has gained prominence in recent years, with scholars like Floridi (2020) exploring the implications of machines as moral agents. Questions of accountability, transparency, bias, and power dynamics are central to understanding and navigating the complex landscape of human-machine collaboration.

WHAT IS COLLABORATIVE AI REALLY ABOUT?

In essence, collaborative AI represents a shift from AI as a stand-alone entity to AI as an integrated and interactive partner, working alongside humans in various domains, from healthcare and education to business and entertainment. Hence, the title *Converging Minds*. What does it mean when minds converge? What we are talking about is a synergistic melding of human thought processes and AI capabilities, leading to a harmonized approach to problem-solving and decision-making. It encompasses the idea that as humans and AI systems work together, they can achieve outcomes that neither could accomplish alone. The convergence can be seen in collaborative workflows, decision-making processes, and creative endeavors, where human intuition and AI's data-driven insights come together, producing enhanced results. In the spirit of this convergence, rather than positioning humans and machines in competition, collaborative AI seeks to forge a partnership that leverages the unique strengths of both. While machines can perform certain tasks with a level of efficiency and precision beyond human capability, they lack the creativity, empathy, and intuition that define human intelligence.

We firmly believe that in the workplace of the future, collaborative AI offers a model where humans and machines work in synergy, accomplishing what neither could achieve alone. Education, medicine, research, and the creative arts are just a few areas where this partnership between humans and machines is beginning to bear fruit. In the field of education, collaborative AI offers the possibility of personalized learning paths that adapt to individual needs and preferences, allowing educators to concentrate on fostering the creative and critical thinking skills that machines cannot replicate. Similarly, in medicine, AI's ability to assist in diagnostics and treatment planning frees physicians to engage more deeply with their patients, providing a level of personalized care that was previously unattainable.

Researchers, too, are finding that collaborative AI can support them in their work, enabling faster data analysis and predictive modeling and allowing them to apply their domain expertise in more innovative and insightful ways. In the creative arts, artists are exploring how AI can enhance their process, blending human intuition and emotion with machine-generated patterns and insights to create works that are both novel and resonant. However, the integration of collaborative AI into these and other fields is not without challenges. There are valid concerns about privacy, security, and the ethical implications of machines taking on roles traditionally filled by humans. Ensuring that these AI systems operate transparently, that they are explainable, and that they align with human values is paramount. Careful thought must be given to how we navigate these challenges if collaborative AI is to fulfill its promise without unintended negative consequences.

WAYS OF COLLABORATING WITH AI

The collaboration between human and artificial intelligence is an interplay, shaped by the nature of human cognition and the ever-evolving capabilities of AI. This section discusses the possible modalities of this collaboration that are already established or emerging. We traced at least five such modalities that can be currently traced. Quite obviously, we do not have to limit ourselves to those presented. The future of human-AI collaboration can comprise new ways of working together that we don't quite yet see (Malone, 2018; Daugherty & James Wilson, 2018).

PARALLEL WORK

Using parallel work, both human and AI agents work on their tasks independently, yet their efforts are directed towards a shared objective. The human agent might engage in tasks that require creativity, ethical discernment, or a deep understanding of context. In contrast, the AI agent often excels in data analysis, pattern identification, and computational tasks. While they operate separately, their combined efforts are harmonized to achieve the desired outcome.

This mode of collaboration offers the advantage of efficiency, as tasks are executed simultaneously, harnessing the unique strengths of both agents. However, certain challenges exist. Integrating the distinct contributions of both agents demands meticulous coordination and communication. There's a need to ensure that methodologies, interpretations, and assumptions align to produce a coherent result. Ethical considerations also come to the fore in parallel work. Questions of accountability, transparency, and inclusivity arise, especially when determining responsibility for the combined output. The collaboration's success hinges on addressing these ethical concerns, ensuring that decisions are transparent and that diverse perspectives are considered.

Parallel work finds its applications in diverse domains, from scientific endeavors to artistic creations. A case in point is medical research, where clinicians provide patient care insights, and AI algorithms sift through vast datasets to unearth trends. Their combined efforts enhance medical practices, from diagnosis to treatment.

Looking ahead, the future of parallel work will be shaped by the development of tools and frameworks that promote effective collaboration. This encompasses mechanisms for integration, ethical guidelines, and systems that can adapt to the dynamic nature of collaboration.

CONTRIBUTIVE WORK

Contributive work stands as a distinctive mode of collaboration, where artificial intelligence functions primarily to augment and enhance human capabilities. In this framework, AI is not an independent agent but a supportive tool, tailored to provide specific functionalities that bolster human creativity, decision-making, and problem-solving. The essence of contributive work lies in its human-centered approach, ensuring that AI's contributions are aligned with the needs, goals, and contexts of human agents. This fosters a collaboration that respects human autonomy and expertise.

The flexibility inherent in contributive work allows AI to be customized to various tasks and domains, from scientific endeavors to artistic pursuits. Such adaptability ensures that AI's contributions are relevant and aligned with the specific objectives of the collaboration. However, this mode of collaboration is not without its complexities. Ethical considerations, particularly those related to transparency, accountability, and potential biases, are paramount. It's vital to ensure that AI's contributions are ethically sound, transparent, and devoid of biases. Moreover, the design and implementation of AI systems in this framework demand meticulous attention, ensuring they are intuitive, responsive, and cognitively aligned with human agents.

Diverse applications of contributive work can be observed across sectors, exemplified in areas like education, where AI supports human educators by personalizing learning materials or providing real-time feedback. As we look to the future, the trajectory of contributive work will be shaped by advancements in AI systems that are more empathetic, context-aware, and ethically aligned. Innovations in human-AI interaction, ethical design principles, and domain-specific customizations will further refine this collaboration modality.

In essence, contributive work underscores the potential of a collaboration that values human expertise while leveraging AI's specific strengths. It offers insights into designing and implementing AI in ways that are supportive, ethically responsible, and harmoniously aligned with human values and aspirations.

TEAMWORK

Teamwork in human-AI collaboration reimagines the traditional understanding of teams. In this modality, artificial intelligence is not only an auxiliary tool or an independent entity but also an integral member of a cohesive team, collaborating side by side with human agents. This conceptualization underscores a partnership where both human and AI agents share responsibilities, engage in effective communication, and adapt to each other's actions and needs. Such a collaborative environment necessitates AI systems that are adept at understanding and responding to human cues, mirroring the dynamics of traditional human teamwork.

The essence of team work lies in its emphasis on shared goals and joint responsibility for outcomes. Clear definitions of roles, expectations, and contributions are pivotal, ensuring alignment and engagement in the collaborative process. Effective communication becomes the linchpin of this collaboration, demanding AI systems that can interpret human instructions, feedback, and even emotions, and respond in contextually appropriate ways. This dynamic interaction fosters enhanced creativity and problem-solving.

Trust emerges as a cornerstone of team work. The mutual reliance between human and AI agents necessitates that AI performs reliably, while also being designed to respect and adapt to human values and preferences. Ethical considerations, particularly those related to autonomy, accountability, and transparency, are also very important. Striking a balance between AI's autonomy and human oversight, coupled with clear accountability mechanisms and algorithmic transparency, is essential for an ethically sound team collaboration.

The complexities of designing AI systems for genuine teamwork cannot be understated. It demands algorithms that not only compute natural language and emotions but also interact in ways that are intuitive and resonate with humanlike nuances. Integrating such AI into existing team dynamics presents its own set of challenges, yet the potential applications, be it in healthcare, research, or entertainment, are vast and transformative.

HYBRID MODELS

In addition to these distinct modalities, there are hybrid models that combine elements of parallel work, contributive work, team work, and synergy that we will describe further on. These models recognize the fluid and context-dependent nature of collaboration, allowing for flexibility and adaptability in response to specific tasks, goals, and environments (Neumayr et al., 2021). The hybrid mode of collaboration with generative AI refers to a working relationship where both human and AI agents contribute to the creative or problem-solving process. In this model, the AI system (or a couple of AI systems/agents) generates ideas, solutions, or content based on its training data and algorithms, while the human collaborator (or a team of collaborators) brings in contextual understanding, ethical considerations, and creative direction that the AI lacks. Generative AI can produce multiple options or pathways, which the human can then refine, select, or combine to achieve the desired outcome. This collaborative approach leverages the strengths of both human intuition and AI's data-processing capabilities, aiming for results that neither could achieve alone.

Hybrid models of collaboration show the convergence of diverse modalities, encompassing elements of parallel work, synergy, contributive work, and teamwork. These models underscore the fluid and context-sensitive nature of human-AI interactions, allowing for a dynamic interplay tailored to specific tasks, goals, and environments. Embracing the fluidity of collaboration, hybrid models facilitate a tailored approach, adapting to the unique requirements of each collaborative endeavor. This adaptability extends to various domains, from research projects that necessitate both parallel exploration and team-based discussions to business strategies that intertwine contributive assistance with synergistic innovation. The inherent complexity of these models stems from the challenge of integrating diverse collaborative modalities into a coherent whole. This integration demands a profound grasp of human cognition, the capabilities of AI, and the dynamics that underpin collaboration.

Ethical considerations remain paramount in the hybrid models. Questions of transparency, accountability, autonomy, and inclusivity permeate these collaborative endeavors. Striking a balance between the multifarious roles of human and AI agents, upholding ethical standards across diverse modes of collaboration, and championing inclusivity are essential facets of these models.

The application of hybrid models spans a wide array of sectors, from healthcare and education to finance and the creative arts. In contexts such as urban planning, these models might manifest as a blend of parallel data analysis, contributive simulations, team-based deliberations, and synergistic design innovations. While the implementation of hybrid models presents its own set of challenges, particularly in ensuring coherence and fostering trust, it also unveils opportunities for groundbreaking collaboration that harnesses the collective strengths of both human and AI agents.

Looking ahead, the trajectory of hybrid models is poised to be shaped by advancements in adaptive AI systems, the evolution of ethical frameworks, and the establishment of best practices that champion effective collaboration. In essence, hybrid models present a multifaceted approach to human-AI collaboration, underscoring the versatility and potential of melding human and machine intelligence in contextually relevant, ethically sound, and creatively enriching ways.

SYNERGY

Synergistic collaboration goes beyond mere parallelism, fostering a dynamic interplay between human and AI agents. In this modality, the interaction is continuous and iterative, with each agent influencing and being influenced by the other. The result is a creative process that is greater than the sum of its parts, where new insights and solutions emerge from the reciprocal relationship between human intuition and AI's data-driven analysis. Synergy in human-AI collaboration represents a dynamic and reciprocal relationship where the interaction between human and AI agents leads to outcomes that are greater than the sum of their individual contributions. Unlike parallel or contributive work, synergy emphasizes continuous and iterative collaboration. This section looks into the characteristics, dynamics, benefits, and challenges of synergistic collaboration.

Synergistic collaboration encapsulates a dynamic interplay between human and AI agents, characterized by continuous and iterative interactions. In this mode of collaboration, each agent not only influences but is reciprocally influenced by the other, fostering a fluidity that amplifies creativity, innovation, and problem-solving capacities. This dynamic is further enriched by the emergence of novel insights, solutions, and perspectives, outcomes that transcend what might be achievable through mere additive collaboration. By harnessing the complementary strengths of human intuition and AI's data-driven prowess, synergy births outcomes that are both novel and impactful.

The distinctive features of synergistic collaboration are underscored by its inherent complexity. Achieving a seamless and productive interaction demands an integration of human cognition with AI algorithms, necessitating a profound grasp of the collaborative dynamics at play. The ethical dimensions of synergy are further intertwined with the imperatives of effective communication and trust. For synergy to truly flourish, human agents must harbor trust in AI's responsiveness, and AI systems must be adept at interpreting human cues with precision.

The applicability of synergy spans a diverse array of domains, from scientific endeavors and business innovations to artistic ventures and the formulation of social policies. An illustrative example can be found in the area of product design, where the confluence of human creativity and AI's simulation capabilities can lead to the conception of innovative and sustainable products.

Yet, like all collaborative modalities, synergy presents its own set of challenges. These range from ensuring coherence and adaptability to navigating the ethical and technological challenges inherent in such collaboration. Despite such challenges, the potential rewards are immense, offering avenues for groundbreaking collaborations that tap into the full potential of both human and AI agents.

Looking to the horizon, the future trajectory of synergy is poised to be shaped by advancements in AI's empathetic capabilities, its ability to understand context, and its alignment with ethical principles. Continued research in areas such as human-AI interaction, natural language processing, and ethical design will undoubtedly influence the evolution and refinement of synergistic collaboration (Figure 0.1).

As you can see on the mind map below, we can segment collaborative AI into five key modes, but also look at their sub-modes. In parallel work, humans and AI operate in silos, focusing on individual tasks with minimal interaction. Contributive work involves task delegation between the two entities but maintains a low level of engagement. Teamwork signifies a high level of interaction, characterized by joint decision-making processes. Hybrid collaboration features mixed teams of humans and AI, with varying degrees of interaction based on task requirements. Lastly, synergy represents the pinnacle of collaboration, where humans and AI are seamlessly integrated, forming a symbiotic relationship that enhances the capabilities of both.

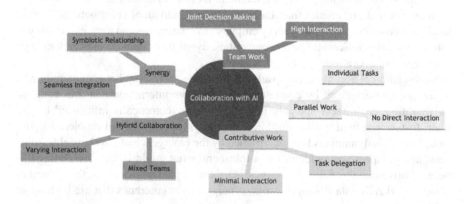

FIGURE 0.1 Mind map diagram illustrating different modes of collaboration with AI generated using Diagrams Plugin within ChatGPT.

GENERATIVE AI'S ROLES AND MODALITIES OF COLLABORATION

As you will see throughout this book, the roles assigned to AI systems can be seen as reflections of the diverse modalities of collaboration. These roles not only define the nature of the interaction but also shape the outcomes and the depth of the collaboration (Hart-Davis, 2023; Franklin, 2022).

The role of a **critic** mirrors aspects of the contributive work modality. As a critic, AI (and in our context ChatGPT or Claude) provides feedback, highlights potential pitfalls, and offers suggestions for improvement. Much like in contributive work, where AI augments human capabilities, the critic role is about enhancing the quality of human outputs by providing targeted assistance. It's a supportive role, ensuring that the final product is refined and free from oversights.

The **style amplifier** role resonates with the synergistic collaboration modality. In synergy, the combined efforts of human and AI lead to emergent outcomes that neither could achieve independently. As a style amplifier, ChatGPT takes the essence of a given content and enhances its presentation, tone, or style, creating a result that is more than just the sum of its parts. It's a dynamic interplay, where the initial human input and AI's stylistic enhancement feed into each other.

The **idea generator** role aligns with aspects of both parallel work and teamwork modalities. In parallel work, both human and AI operate independently but towards a shared goal. As an idea generator, ChatGPT independently comes up with concepts, themes, or solutions that can then be integrated with human ideas. In the context of teamwork, the idea generation becomes a more interactive process, where human and AI brainstorm, communicate, and build upon each other's suggestions, much like members of a cohesive team.

Lastly, the **field expert** role embodies the hybrid models of collaboration. Being a field expert requires a fusion of various capabilities: providing in-depth knowledge (akin to contributive work), interacting and adapting to user queries (reminiscent of teamwork), and generating insights or solutions (paralleling both synergy and parallel work). This role recognizes the fluid and context-dependent nature of collaboration, adapting to the specific needs of the user and the topic at hand.

The roles assigned to AI systems, such as critic, style amplifier, idea generator, and field expert, are functional and also emblematic of the broader modalities of human-AI collaboration. They showcase the potential, adaptability, and depth of interactions between humans and machines, paving the way for a future where AI is a genuine collaborator.

NEW ROLES AND SKILLS FOR HUMANS IN A COLLABORATIVE AI ECOSYSTEM

We can assign different roles to AI, but how about the human-side of the equation? Surely, AI should be human-centric and not the other way around, but the dawn of the collaborative AI ecosystem brings novel roles and demands specialized skills, and thus reshapes our societal and professional roles as well. As we stand at the intersection of human expertise and machine intelligence, understanding these new dynamics is crucial for our preparedness in the evolving world of work (James Wilson et al., 2017; Lee & Qiufan, 2021). We will now look into some emerging roles and professions that come together with the AI revolution.

At the forefront, the role of an AI trainer emerges as potentially paramount (Moore, 2020). Unlike traditional programming, training an AI model demands a deep understanding of data nuances, biases, and real-world contexts. AI trainers imbue models with sensitivity to these aspects, ensuring their outputs align with human values and ethical standards. This role is less about coding prowess and more about insightful guidance, making it a blend of technical acuity and sociocultural understanding.

Parallelly, as AI systems become more embedded in our daily lives, there is an observable rising demand for AI ethicists (Gambelin, 2021). Their responsibility extends beyond compliance. AI ethicists navigate the nebulous territories of moral dilemmas and societal impacts, offering guidelines to ensure AI's alignment with human-centric values. Grounded in philosophy, sociology, and technology, their expertise helps in the harmonious integration of AI systems into society without infringing on individual rights or perpetuating biases.

With AI systems often interacting directly with users, the role of a human-AI interaction designer slowly gains prominence (Kore, 2022). Their expertise revolves around creating intuitive interfaces, ensuring that AI systems are functional and user-friendly. Their challenge lies in making sophisticated technology accessible and understandable to a diverse user base, bridging the gap between complex algorithms and everyday user experiences.

The collaborative AI ecosystem also sees the evolution of traditional roles. For instance, some educators may soon find themselves transitioning into AI-enhanced learning facilitators (Kerres & Buntins, 2020). Rather than just imparting knowledge, they leverage AI tools to curate personalized learning experiences, identify individual student needs, and optimize educational pathways. Their role becomes more about guiding and mentoring, fostering environments where AI assists in tailored educational delivery.

The therapeutic area is also experiencing an AI-infused renaissance. For example, the AI therapeutic counselors are professionals who leverage AI tools to provide mental health support (Quinn, 2023). While the core of therapy remains human empathy, these counselors utilize AI-driven insights to identify patterns, predict potential stressors, and tailor interventions. Their expertise extends beyond traditional therapeutic methods, merging cognitive techniques with real-time data to offer holistic mental well-being strategies.

On the creative spectrum, the role of AI-assisted content curators (Kumar et al., 2023) is gaining traction. Beyond the domain of algorithm-driven recommendations, these individuals harness AI's predictive capabilities to offer bespoke cultural and entertainment experiences. They navigate the vast ocean of digital content, sifting through vast data points to present audiences with experiences uniquely tailored to their evolving tastes and preferences. This role epitomizes the symbiosis of human intuition and AI analytics in curating enriching experiences.

Furthermore, urban planning and infrastructure development witness the rise of AI urban strategists (As et al., 2022). Tasked with designing future-ready cities, these strategists harness AI insights to predict population movements, resource consumption, and environmental impacts. Their role is instrumental in sculpting urban landscapes that are sustainable and adaptable, catering to the fluctuating needs of modern societies while ensuring ecological balance.

In the financial sector, new roles may include AI financial ethnographers who could form a professional group studying the relationship between AI-driven financial systems and diverse communities (Tett, 2021). They analyze how different communities interact with, benefit from, or are marginalized by automated financial operations, offering insights to make these systems more inclusive and culturally sensitive.

Another type of a new profession is the AI transparency auditor (Spair, 2023). These individuals are tasked with deciphering the often opaque decision-making processes of AI systems. Their main objective is not just to understand but to elucidate the AI models for the broader public, ensuring accountability and trustworthiness. Rooted in both technological expertise and effective communication, they demystify AI's complex operations, making its decisions more comprehensible and justifiable to stakeholders.

However, this evolving ecosystem is not without its challenges. The infusion of AI in various sectors necessitates a continuous upskilling of the workforce. In addition to being proficient in one's core domain, an understanding of AI's capabilities and limitations becomes an essential facet of many professions. Doctors, lawyers, artists, and agriculturists find themselves needing to comprehend and integrate AI insights into their work. Preparing for this future demands a reimagining of education, training, and continuous learning, ensuring that as AI evolves, humanity evolves alongside it, harnessing its potential while retaining the core of what makes us human.

In this complex structure of roles and responsibilities, one overarching theme emerges: the collaborative AI ecosystem doesn't replace human expertise but enhances and refines it. The future of work here is not about competition between man and machine but about collaboration, where each augments the other's strengths. This harmonious integration demands a shift in perspective, emphasizing lifelong learning, adaptability, and an unwavering commitment to ethical considerations. Only through such an approach can we truly harness the boundless potential that the confluence of human and AI capabilities promises.

CONCLUSIONS

With human-AI collaboration, we have traveled through various modalities and roles that shape our interactions with artificial intelligence. These insights illuminate the potential of AI as an active participant in our endeavors.

- Parallel work emphasizes the distinct yet complementary roles of humans and AI, where both entities operate independently but converge towards a shared objective. This approach leverages the unique strengths of human creativity and AI's computational power, and its future is poised to refine integration mechanisms and foster ethical guidelines.

- In contributive work, AI serves as an augmentative tool, enhancing human decision-making and creative exploration. As we look ahead, the trajectory of contributive work is likely to be shaped by AI systems that are increasingly empathetic, context-aware, and ethically aligned.

- Teamwork reimagines the human-AI relationship as a cohesive partnership. Both agents, human and AI, share responsibilities, communicate, and adapt to each other's actions and needs. The promise of teamwork lies in advancing AI's capabilities in empathy, contextual understanding, and ethical alignment, making AI an even more active and integral team member.

- Synergistic collaboration captures the essence of a dynamic interplay between human and AI agents. This continuous and iterative interaction leads to emergent outcomes, novel insights, and solutions that might not have been possible through independent efforts. The evolution of synergy will undoubtedly be influenced by advancements in AI's empathetic capabilities, its ability to understand context, and its alignment with ethical principles.

- Hybrid models of collaboration represent a fusion of these modalities, recognizing the fluid and context-dependent nature of human-AI interactions. The future of these models lies in the development of adaptive and context-aware AI systems, combined with ethical frameworks and best practices that facilitate effective collaboration (Figure 0.2).

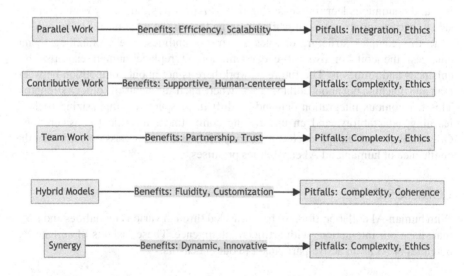

FIGURE 0.2 Modalities of collaboration with AI highlighting benefits and pitfalls generated using Diagrams Plugin within ChatGPT.

As AI continues to evolve, collaboration between humans and AI will become more nuanced, adaptive, and contextually relevant, promising a harmonious, ethical, and innovative future for human-machine interactions. This symbiotic relationship, rooted in collaboration, holds the promise of unlocking unprecedented avenues of creativity, innovation, and progress for the future. Understanding more closely and granularly the ways we can collaborate with AI is the reason we wrote this book.

REFERENCES

Anderson, J. L., & Coveyduc, J. L. (2020). *Artificial Intelligence for Business: A Roadmap for Getting Started with AI.* John Wiley & Sons.

As, I., Basu, P., & Talwar, P. (2022). *Artificial Intelligence in Urban Planning and Design: Technologies, Implementation, and Impacts.* Elsevier.

Baxter, J., & Koehler, M. (2013). *Leadership through Collaboration: Alternatives to the Hierarchy.* Routledge.

Bicchi, A., Scilingo, E. P., Ricciardi, E., & Pietrini, P. (2008). Tactile flow explains haptic counterparts of common visual illusions. *Brain Research Bulletin, 75*(6), 737–741.

Borowik, G., Chaczko, Z., Jacak, W., & Łuba, T. (2015). *Computational Intelligence and Efficiency in Engineering Systems.* Springer.

Bostrom, N. (2014). *Superintelligence: Paths, Dangers, Strategies.* Oxford University Press.

Brynjolfsson, E., & McAfee, A. (2014). *The Second Machine Age: Work, Progress, and Prosperity in a Time of Brilliant Technologies.* W. W. Norton & Company.

Chang, Y., Wang, X., Wang, J., Wu, Y., Yang, L., Zhu, K., Chen, H., Yi, X., Wang, C., Wang, Y., Ye, W., Zhang, Y., Chang, Y., Yu, P. S., Yang, Q., & Xie, X. (2023). A survey on evaluation of large language models. In *arXiv [cs.CL].* arXiv. http://arxiv.org/abs/2307.03109

Coradeschi, C. M. (2013). *The Rainmaker's Quick Guide to Lasting Sales Success.* Thomas Noble Books.

Daugherty, P. R., & James Wilson, H. (2018). *Human + Machine: Reimagining Work in the Age of AI.* Harvard Business Press.

Feldman, S. (salevati). (2017). *Co-Creation: Human and AI Collaboration in Creative Expression.* https://doi.org/10.14236/ewic/eva2017.84

Floridi, L. (2020). AI and its new winter: From myths to realities. *Philosophy & Technology, 33*(1), 1–3.

Franklin, D. (2022). *The Chatbot Revolution: ChatGPT: An In-Depth Exploration.* Amazon Digital Services LLC – Kdp.

Fussell, S. R., & Kreuz, R. J. (2014). *Social and Cognitive Approaches to Interpersonal Communication.* Psychology Press.

Gambelin, O. (2021). Brave: what it means to be an AI Ethicist. *AI and Ethics, 1*(1), 87–91.

Goertzel, B. (2016). *The AGI Revolution: An Inside View of the Rise of Artificial General Intelligence.* Humanity+ Press.

Goertzel, B., Iklé, M., & Potapov, A. (2022). *Artificial General Intelligence: 14th International Conference, AGI 2021, Palo Alto, CA, USA, October 15–18, 2021, Proceedings.* Springer Nature.

Hargreaves, A., & Fullan, M. (2015). *Professional Capital: Transforming Teaching in Every School.* Teachers College Press.

Hart-Davis, G. (2023). *Killer ChatGPT Prompts: Harness the Power of AI for Success and Profit.* John Wiley & Sons.

Hutter, M. (2005). *Universal Artificial Intelligence: Sequential Decisions Based on Algorithmic Probability.* Springer Science & Business Media.

James Wilson, H., Daugherty, P. R., & Morini-Bianzino, N. (2017). *The Jobs That Artificial Intelligence Will Create: A Global Study Finds Several New Categories of Human Jobs Emerging, Requiring Skills and Training That Will Take Many Companies by Surprise.* MIT Sloan Management Review.

Jemielniak, D. (2014). *Common Knowledge?: An Ethnography of Wikipedia.* Stanford University Press.

Jemielniak, D., & Przegalinska, A. (2019). *Collaborative Society.* MIT Press.

Kerres, M., & Buntins, K. (2020). Recommender in AI-enhanced learning: An assessment from the perspective of instructional design. *Open Education Studies, 2*(1), 101–111.

Koch, J., & Oulasvirta, A. (2018). Group cognition and collaborative AI. In J. Zhou & F. Chen (Eds.), *Human and Machine Learning: Visible, Explainable, Trustworthy and Transparent* (pp. 293–312). Springer International Publishing.

Kore, A. (2022). *Designing Human-Centric AI Experiences: Applied UX Design for Artificial Intelligence.* Apress.

Koschmann, T. (2012). *Cscl: Theory and Practice of An Emerging Paradigm.* Routledge.

Kumar, A., Nayyar, A., Sachan, R. K., & Jain, R. (2023). *AI-Assisted Special Education for Students with Exceptional Needs.* IGI Global.

Kumar, & Muneesh. (2011). *Trust and Technology in B2B E-Commerce: Practices and Strategies for Assurance: Practices and Strategies for Assurance.* IGI Global.

Kurzweil, R. (n.d.). *The singularity is near: when humans transcend biology.* Viking. New York.

Lee, K.-F., & Qiufan, C. (2021). *AI 2041: Ten Visions for Our Future.* Crown.

Loos, T. B. (2023). *AI & AGI – The Convergence of Mind and Machine.* T.B.I Publishing

Ludik, J. (2021). *Democratizing Artificial Intelligence to Benefit Everyone: Shaping a Better Future in the Smart Technology Era.* Amazon Digital Services LLC – KDP Print US.

Lukowicz, P., Mayer, S., & Koch, J. (2023). *HHAI 2023: Augmenting Human Intellect: Proceedings of the Second International Conference on Hybrid Human-Artificial Intelligence.* IOS Press.

Mahajan, P. (2021). *Artificial Intelligence in Healthcare: AI, Machine Learning, and Deep and Intelligent Medicine Simplified for Everyone.* MedMantra, LLC.

Malone, T. W. (2018). *Superminds: The Surprising Power of People and Computers Thinking Together.* Little, Brown.

McLuhan, M., Terrence Gordon, W., Lamberti, E., & Scheffel-Dunand, D. (2011). *The Gutenberg Galaxy: The Making of Typographic Man.* University of Toronto Press.

Moore, P. V. (2020). *AI Trainers: Who Is the Smart Worker Today?* https://papers.ssrn.com/abstract=3883910

Naughton, J. (2014). *From Gutenberg to Zuckerberg: Disruptive Innovation in the Age of the Internet.* Quercus.

Neumayr, T., Saatci, B., Rintel, S., Klokmose, C. N., & Augstein, M. (2021). What was hybrid? A systematic review of hybrid collaboration and meetings research. In *arXiv [cs.HC].* arXiv. http://arxiv.org/abs/2111.06172

OpenAI (2023). *Pioneering research on the path to AGI.* Retrieved from https://openai.com/research/overview

Quinn, H. (2023). *AI and Mental Health: Transforming Therapy and Counseling.* Amazon Digital Services LLC – Kdp.

Russell, S. J., & Norvig, P. (2016). *Artificial Intelligence: A Modern Approach.* Malaysia; Pearson Education Limited.

Sabry, F. (2023). *Narrow Artificial Intelligence: Fundamentals and Applications.* One Billion Knowledgeable.

Schmidhuber, J., Thorisson, K. R., & Looks, M. (2011). *Artificial General Intelligence: 4th International Conference, AGI 2011, Mountain View, CA, USA, August 3–6, 2011, Proceedings.* Springer.

Shneiderman, B. (2022). *Human-Centered AI*. Oxford University Press.

Sowa, K., Przegalinska, A., & Ciechanowski, L. (2021). Cobots in knowledge work: Human – AI collaboration in managerial professions. *Journal of Business Research*, *125*, 135–142.

Spair, R. (2023). *AI in Practice: A Comprehensive Guide to Leveraging Artificial Intelligence*. Rick Spair.

Stahl, G. (2010). *Global Introduction to CSCL*. Lulu Enterprises Incorporated.

Tett, G. (2021). *Anthro-Vision: A New Way to See in Business and Life*. Simon and Schuster.

Vygotsky, L. S. (1980). *Mind in Society: The Development of Higher Psychological Processes*. Harvard University Press.

1 AI in Research, Innovation and Education

This chapter dives deep into the origins of chatbots, tracing their evolution and impact on various fields. We explore how these early conversational agents have influenced art, serving as both medium and muse for artists. In the educational sector, we examine the chatbots' role in personalized learning and student engagement. The chapter sets the stage for understanding the broader implications of AI, offering a lens through which to view its transformative potential. It serves as a foundation for the subsequent exploration of more advanced forms of human-AI collaboration.

The history of artificial intelligence is marked by many chapters, each highlighting a different capability. Among these, generative AI represents a transformative leap, not due to its technical innovation but primarily because of its capacity to augment and expand human creativity and problem-solving. Generative AI encompasses a set of technologies that can generate novel data patterns, be it in the form of text, images, music, or complex simulations. What sets it apart is its ability to produce outputs that are not explicitly programmed into it, drawing inspiration from vast datasets and, in some instances, mirroring the unpredictability of human creativity (Du Sautoy, 2020; Hageback, 2021).

In practical domains, there is a significant augmentation potential of generative AI. In design and architecture, professionals can employ generative models to quickly visualize potential designs enabling them to assess and choose options that may not have been conceived through conventional means. In music, artists can use these tools to explore unique compositions, merging traditional musical frameworks with AI-generated innovations. In the area of scientific research, generative AI can be employed to predict molecular structures, simulate environmental changes, and model the potential spread of diseases. These models can provide rapid insights, thereby accelerating the pace of innovation and discovery (Abdulkareem & Petersen, 2021; Tsigelny, 2019).

Moreover, generative models, with their capacity to rapidly analyze and predict novel outcomes, have the potential to significantly accelerate the trajectory of discoveries, bridging gaps between theoretical exploration and practical solutions.

However, as the lines between AI-generated and human-created content become more fluid, ethical and philosophical questions arise. The essence of originality, the nature of authenticity, and the definition of intellectual property are placed under scrutiny. In a world where a piece of art or a groundbreaking solution can be co-created with AI, society must grapple with the shifting sands of ownership and value attribution.

Equally pressing are the challenges posed by the unpredictable nature of generative outputs. The same capacity that enables generative AI to produce groundbreaking

DOI: 10.1201/9781032656618-2

solutions can, without careful oversight, lead to misleading and detrimental results (Yu et al., 2023). The emergence of deepfakes, which harness generative technology to create disturbingly realistic forgeries, stands as a reminder of the potential risks. But we will get back to that later.

In this part of the chapter, we venture into the heart of the dynamic intersection of generative AI in augmenting human capabilities. We will also explore the moral, ethical, and societal quandaries it presents. Through thoughtful exploration, we will look into the transformative potential of AI in augmenting human creativity, scientific inquiry, as well as reshaping content generation (Carter & Nielsen, 2017). We invite you to envision a future where AI is a partner, shaping the future of creative expression in ways we are only beginning to comprehend. Historically, creativity was seen as the exclusive domain of humans, an indication of our unique ability to imagine, innovate, and inspire. However, with the advent of advanced, generative AI systems, we are witnessing an interesting paradigm shift. AI systems, equipped with vast datasets and sophisticated algorithms, can now produce content that rivals, and in some cases surpasses, human-generated works in terms of complexity and aesthetic appeal. Rivalry, however, is not the direction. As we have underlined before, the intersection of human creativity and artificial intelligence is the exciting frontier, teeming with potential and full of possibilities. This chapter seeks to unravel the distinctive features of this union, exploring the implications of collaborative AI in the world of creative work, research, and content generation (Harris & Waggoner, 2019). But let us start with a story.

WHO OR WHAT IS CHATGPT AND WHERE DID IT COME FROM?

In the field of artificial intelligence and machine learning, there are few innovations that have had the lasting impact of chatbots. While we often consider them a product of our modern technological era, the concept of a machine capable of simulating human conversation has roots that stretch back over half a century.

So, our story begins in 1966 at the Massachusetts Institute of Technology (MIT), with a revolutionary program named ELIZA. Conceived by computer scientist Joseph Weizenbaum, ELIZA was a remarkable leap in natural language processing and computational linguistics. What set ELIZA apart wasn't just the ability to process user input and generate responses, but the uncanny skill it demonstrated in mimicking the back-and-forth rhythm of human conversation.

ELIZA was designed to imitate a Rogerian psychotherapist, using a technique called reflection to turn the user's statements back as questions. For instance, if a user were to tell ELIZA, "I'm feeling a bit stressed," ELIZA might respond, "Why do you say that you're feeling a bit stressed?" While this technique was relatively simple, it gave the illusion of comprehension and empathy, causing some users to believe they were conversing with a human.

The creation of ELIZA marked a shift, demonstrating the potential of computers to interact with humans in a more natural and intuitive way. However, despite the groundbreaking design, ELIZA was far from perfect. There was no understanding or awareness of the content of the conversations. Instead, the responses were entirely rules-based, reflecting the patterns programmed by its creator.

Nonetheless, ELIZA laid the groundwork for the sophisticated chatbots we see today, which employ advanced machine learning algorithms to process and understand human language with ever-increasing accuracy. As we delve deeper into this topic, we will appreciate the significant strides made since ELIZA's time and explore the exciting (and challenging) implications of the current state of chatbot technology.

Building on the foundational work initiated by ELIZA, the field of conversational AI has advanced significantly over the decades, culminating in the development of chatbots with far greater capability and sophistication.

THE RISE OF RULE-BASED SYSTEMS AND MACHINE LEARNING

Following ELIZA, the next wave of conversational systems was dominated by rule-based chatbots. These bots operated based on a set of predefined rules and decision trees. They could handle more complex interactions than ELIZA but were limited by the rules they were programmed with. Any deviation from expected inputs would often stump these systems. Further on, as computational power grew and machine learning techniques advanced, chatbots began to evolve. Instead of relying solely on hardcoded rules, they started learning from data. By analyzing large datasets of human conversations, these systems could generate responses based on patterns and probabilities. This shift marked the beginning of a new era for conversational AI, allowing for more flexibility and adaptability.

However, the true transformation in conversational systems came with the advent of deep learning. Models like ChatGPT, built on the GPT (Generative Pre-trained Transformer) architecture, leveraged neural networks with millions of parameters. Trained on diverse and extensive textual data, these models could generate coherent, contextually relevant, and often indistinguishably humanlike responses. ChatGPT, developed by OpenAI, represented a significant leap in this domain. Not only it could engage in fluid conversations, but it could also understand context, generate creative content, and adapt to specific user instructions. The line between human and machine communication began to blur, opening up tons of possibilities for applications ranging from customer support to creative writing assistance. Unlike ELIZA, which operated purely on rules, ChatGPT is trained on large amounts of text data, enabling it to generate contextually relevant responses and display a form of creativity. Although ChatGPT does not truly understand or have consciousness, its mimicry of humanlike conversation represents a significant advancement in natural language processing.

REFLECTING ON THE JOURNEY

From the rudimentary reflections of ELIZA to the nuanced interactions of ChatGPT, the evolution of conversational systems has been a proof of human creativity and technological advancement. These systems have transitioned from simple rule-followers to adaptive learners, mirroring the complexities of human dialogue. This progression demonstrates the evolving relationship of our society with machines. It is clear that the legacy of early conversational systems like ELIZA has blossomed into a dynamic partnership between humans and machines, reshaping the cultural narrative for the future. We have advanced from basic conversational models to sophisticated AI collaborators, and we've also witnessed a deep impact on our cultural fabric. Culture has always been a reflection of societal advancements. As technology has evolved, so too has its influence on cultural mediums. The introduction of ELIZA in the 1960s marked a seminal moment, showcasing the potential of machines to engage in rudimentary dialogue (Weizenbaum, 1966; Weizenbaum & McCarthy, 1977). But the journey from ELIZA's basic conversational capabilities to today's generative AI models represents more than just technological progress; it signifies a transformative shift in how we collaborate with machines in the domain of culture (Monaco & Woolley, 2022).

As we underlined a couple of times, the role of AI is not to replace but to augment. It acts as a collaborator, enhancing the creative process, offering new perspectives, and opening doors to uncharted artistic territories. In the digital age, the boundaries between human creativity and machine-generated content have become increasingly blurred. With advancements in natural language processing, understanding, and generation, the next generation of chatbots might be indistinguishable from human interlocutors (Bryson et al., 2017). As we continue to push the boundaries of what's possible, the story of conversational systems serves as a reminder of how far we've come and the limitless horizons that lie ahead.

Collaborative AI, as we define it, is not about replacing the human creator but augmenting their capabilities. It's about a partnership where each entity brings its unique strengths to the table. While humans possess intuition, emotion, and rich lived experiences, AI offers enormous computational power, pattern recognition, and the ability to process and generate content at a large scale. By harnessing the strengths of both human creators and AI tools, we can achieve a synergy that elevates the creative process to unprecedented heights. From generating novel ideas and refining artistic styles to critiquing and enhancing content, collaborative AI emerges as a powerful ally in the creative journey.

THE SYNERGY OF EMOTION AND LOGIC

The synergy between human emotion and machine logic forms human-AI collaboration.While AI excels in analyzing data, identifying patterns, and generating content based on predefined parameters, it lacks the emotional depth and subjective experiences that humans have. This emotional depth, rooted in our personal experiences, cultural backgrounds, and individual perspectives, adds a layer of richness and nuance to the creative process. When combined with the precision and efficiency of AI, the result is a blend of emotion and logic, as well as intuition and analysis.

For example, in visual arts, Fink and Akdag Salah (2023) have been experimenting with AI tools to create mesmerizing artworks that blur the lines between man and machine. In one notable instance, an AI-generated artwork that we will mention later was auctioned at Christie's for a staggering sum, signaling the art world's recognition of *machine-generated creativity*. Behind this artwork, however, was not only an algorithm but also the culmination of human-artistic intent and machine precision. The artist provided the initial input, the vision, and the AI extrapolated, refined, and brought that vision to life in unexpected ways.

As with any technological shift, the rise of AI in creative work brings ethical implications. Issues of copyright, intellectual property rights, and the very essence of artistic ownership are now under the spotlight. If an artwork is co-created by an artist and an AI, who holds the rights to it? What percentage of the creation can be attributed to the machine, and what to the human? Furthermore, as AI systems become more sophisticated, there's a looming fear of machines overshadowing human creators. Will there come a day when AI-generated content is indistinguishable from human-generated content? And if so, what place will human creators hold in such a landscape?

"EDMOND DE BELAMY": THE AI-GENERATED PORTRAIT

In 2018, the art world was taken by storm with the unveiling of "Edmond de Belamy," a portrait not crafted by the hand of a human artist but generated by a machine learning algorithm. This artwork, created by the Paris-based art collective Obvious, became a symbol of the intersection of art, technology, and culture (Goenaga, 2020; Stephensen, 2019).

The portrait was produced using a generative adversarial network (GAN), a type of machine learning model that we will tell you more about in the next chapter. For now, it is important to know that GANs consist of two parts: a generator, which creates images, and a discriminator, which evaluates them. The model was trained on a dataset of 15,000 portraits painted between the 14th and 20th centuries. As the GAN processed this vast collection, it began to understand the nuances, styles, and common features of these historical portraits. The generator then started producing its own images, attempting to mimic the styles it had learned. Each generated image was critiqued by the discriminator. Through this iterative process of creation and critique, the GAN refined its outputs until it produced the final portrait of "Edmond de Belamy."

The portrait, with its blurry features and dreamlike quality, evoked a sense of mystery and intrigue. It raised questions about authorship, creativity, and the role of machines in the artistic process. When "Edmond de Belamy" was auctioned at Christie's, it fetched an astounding $432,500, *far surpassing its estimated value*. This event marked a significant moment in the art world, signaling a growing acceptance and curiosity about AI-generated art.

The artwork also sparked debates among artists, technologists, and philosophers. Some praised the innovative use of technology, seeing it as a new frontier in artistic expression. Others expressed concerns about authenticity and the potential devaluation of human creativity.

The success of "Edmond de Belamy" opened the doors for further exploration of AI in art. Artists worldwide began experimenting with GANs and other AI models, producing artworks that ranged from abstract paintings to sophisticated sculptures.

Beyond just art, the portrait's impact resonated in discussions about the broader cultural implications of AI. It became a touchstone for debates on AI's role in creative industries, from music and literature to film and theater.

SUNSPRING: THE AI-WRITTEN SCI-FI SHORT FILM

In 2016, the world of cinema witnessed a unique experiment: a short film titled *Sunspring*, where the screenplay was entirely written by an artificial intelligence. Directed by Oscar Sharp and starring Thomas Middleditch, this film showcased the potential and peculiarities of AI-driven storytelling (Cohn, 2021; Riedl et al., 2011).

The AI behind *Sunspring* was named Benjamin, a recurrent neural network trained on a multitude of sci-fi screenplays. Using this extensive training data, Benjamin was tasked with crafting a screenplay for a short film. The result was a script that, while structurally coherent, was filled with unexpected dialogues, unconventional character interactions, and a narrative that defied traditional storytelling norms. The team decided to produce the film exactly as the AI wrote it without any human edits to the script. This decision led to a film that was both intriguing and surreal, with characters uttering lines that were at times profound and at other times nonsensical. *Sunspring* was met with a mix of amusement, bewilderment, and admiration. While the narrative was undeniably disjointed, the film offered a fascinating glimpse into the AI's interpretation of human emotions, relationships, and existential dilemmas.

The film sparked discussions about the nature of creativity. Was *Sunspring* a genuine piece of art, or was it just a technological novelty? The film also raised questions about the future of the entertainment industry. Could AI one day replace human screenwriters, or would it serve as a tool to augment human creativity? Rest assured, the experiment with *Sunspring* led to further explorations of AI in filmmaking. While Benjamin's narrative was abstract and unconventional, it opened the door for filmmakers to consider AI as a collaborative partner in the creative process. This collaboration could range from brainstorming plot ideas to generating dialogues or even predicting audience reactions.

Beyond filmmaking, *Sunspring* became a point of reference in discussions about AI's role in other creative fields, such as literature, music, and theater. It served as both a cautionary tale and an inspiration, highlighting the potential and pitfalls of AI-driven creativity. *Sunspring* stands as evidence of the unpredictable and boundless nature of AI-driven art. It challenges us to reconsider our definitions of narrative, creativity, and artistry. In a world where machines can craft stories, *Sunspring* prompts us to reflect on the essence of human storytelling and the future of collaborative creation in the age of AI.

"HELLO WORLD": THE FIRST AI-COMPOSED ALBUM

In music, composers collaborate with AI to explore new melodies, rhythms, and harmonies, pushing the boundaries of existing norms. In literature, writers can use AI to brainstorm plot twists, develop characters, andcraft poetry that resonates with a global audience. Visual artists can use AI tools to experiment with styles, mediums, and techniques, creating artworks that are both innovative and evocative. In the ever-evolving landscape of music, 2017 witnessed a groundbreaking moment: the release of "Hello World", an album entirely composed by artificial intelligence. Developed by the French collective Skygge, which translates to "shadow" in Danish, this project was a pioneering exploration of the harmonization of human and machine in musical composition (Seok, 2023).

The AI behind "Hello World" was named Flow Machines. It was developed by Sony's Computer Science Laboratories in Paris and was trained on a vast database of songs from various genres, ranging from jazz and pop to classical compositions. Using this diverse musical knowledge, Flow Machines could generate melodies, harmonies, and even intricate chord progressions. For the "Hello World" project, musicians and artists collaborated with the AI. They provided initial inputs, like a theme or a mood, and Flow Machines generated musical segments in response. These segments were then refined, arranged, and produced by human musicians to create the final tracks for the album.

"Hello World" was met with intrigue and critical acclaim. While some tracks resonated with traditional musical sensibilities, others ventured into experimental territories, offering sounds and harmonies that were fresh and unexpected. The album sparked discussions about the essence of musical creativity. Could a machine capture the emotional depth and nuance inherent in music? Or was it just replicating patterns it had learned? Moreover, "Hello World" raised questions about authorship and originality in the age of AI-driven creation.

The success of "Hello World" inspired musicians globally to experiment with AI tools. From generating background scores for films to crafting intricate beats for electronic music, AI became a new instrument in the musician's toolkit. Beyond just composition, AI tools began to be used for mastering tracks, predicting music trends, and even personalizing music experiences for listeners based on their preferences and moods.

CONTROVERSIES AND OUTLOOK

The integration of generative AI into art has sparked several controversies, reflecting deeper societal and philosophical concerns. First of all, questions of authorship, originality, and authenticity. How do we attribute value in a world where machines play a pivotal role in the creative process? When a piece of art is generated by an AI, who owns it? Is it the developer of the AI, the user who provided the input, or the AI itself? This blurring of lines challenges traditional notions of creativity and originality. What does it mean for a piece of content to be "original" when it's generated by an algorithm? And as AI systems become more adept at generating content, how do we ensure that the human touch, the essence of creativity, is not lost? One of the most debated issues is the question of authorship. For instance, the sale of "Edmond de Belamy" for a significant sum at Christie's did raise eyebrows and questions about the valuation of machine-made art. In a recent (2023) ruling by a U.S. court in Washington, D.C., artworks created solely by artificial intelligence without human involvement are ineligible for copyright protection under U.S. law. U.S. District Judge Beryl Howell emphasized that only creations with human authors can be copyrighted. This decision was in response to an application by computer scientist Stephen Thaler for his AI system, DABUS. Thaler, who has faced similar challenges in obtaining U.S. patents for inventions claimed to be created by DABUS, plans to appeal the decision. The ruling underscores the emerging intellectual property challenges in the rapidly evolving field of generative AI. While artists are increasingly integrating AI into their creative processes, the legal landscape remains uncertain, with several lawsuits pending over the use of copyrighted works to train AI models. Judge Howell acknowledged the complexities introduced by AI in the art domain but maintained that human authorship remains a foundational principle of copyright, rooted in long-standing legal traditions.

The democratization of AI tools in the art world brings forth a complex set of economic and ethical considerations (Luce 2019; Luchs 2023). One of the most pressing concerns is the potential economic displacement of human artists. As AI becomes increasingly proficient in generating art, it can produce works at a speed and scale that human artists cannot match. This efficiency, while impressive, poses a risk of flooding the market with AI-generated art, which could be sold at a fraction of the cost of human-created pieces. The economic implications are far-reaching, affecting not just visual artists but also musicians, writers, and creators in other fields. The fear is that human artists might find it increasingly difficult to compete in a marketplace dominated by quick and inexpensive AI-generated works. *The article by Benedict Evans* explores the evolving landscape of intellectual property in the era of generative AI. It raises pertinent questions about ownership and compensation when AI mimics or generates creative works, be it in music, art, or journalism. While traditional intellectual property laws offer some guidance, the scale and capabilities of AI introduce new complexities. The article suggests that these challenges are not just legal but ethical, necessitating a reevaluation of existing frameworks. For instance, while AI doesn't store specific articles, it does rely on the aggregate of human-created content for training. This raises questions about fair compensation and "fair use" of collective human intelligence.

The article also touches on the future of AI, suggesting that as technology advances, AI models may require less data to produce the same or better results. This could potentially alleviate some intellectual property concerns. Additionally, one should emphasize that these models are tools that can be used to create both art and mundane content. It raises questions about the quality and originality of AI-generated works, especially as they become more prevalent, and how this will impact existing artists and the discovery of new, quality content.

This economic challenge is compounded by ethical questions surrounding the training data used by generative AI models. These models are often trained on vast datasets sourced from the internet, which may include copyrighted material or specific artistic styles. While the AI's output may appear original, it could inadvertently reproduce elements of existing works, raising the specter of unintentional plagiarism. This is a significant concern in the art community, where the originality of expression is highly valued. It also poses legal challenges, as artists whose work has been unintentionally replicated by AI may seek legal recourse, further complicating the landscape.

Moreover, the ethical dimension extends to the question of consent. Many artists share their work online for public viewing but not for training machine learning models. The use of such data without explicit permission raises ethical concerns about data ownership and the rights of artists to control how their work is used.

Critics also argue that art is an expression of human experience, emotion, and perspective. They believe that AI, lacking consciousness and emotion, can only mimic styles and patterns but cannot infuse art with genuine sentiment or soul. This sentiment was also evident in reactions to "Hello World," where some felt the music lacked the depth and nuance of human-composed pieces. The concern over the potential loss of human creativity in the sphere of art due to the increasing involvement of AI is a nuanced and forward-looking argument. While AI offers a plethora of styles and techniques, its capabilities are fundamentally shaped by the data it's trained on. If this data predominantly represents mainstream or popular artistic styles, the AI's output is likely to reflect those biases, leading to a homogenization of art.

The richness of art lies in its diversity, its ability to challenge norms, and its capacity to introduce new perspectives. Art has always been a space for avant-garde movements that push boundaries and provoke thought. However, if AI-generated art becomes overly reliant on popular training data, there's a risk that these fringe or experimental art forms may be marginalized. The AI would be less likely to generate art that deviates from the norm, thus reducing the overall diversity of artistic expression.

Moreover, the use of AI in art creation could inadvertently discourage human artists from taking risks or exploring unconventional paths. Knowing that AI can quickly generate art that caters to popular taste might deter artists from investing time and emotional energy into creating something truly unique or controversial. This could lead to a creative stagnation, where both human and machine-generated art becomes increasingly formulaic and predictable.

Additionally, the algorithms themselves could become gatekeepers of what is considered "good" or "valuable" art. If AI tools are designed to optimize for certain styles or themes that are deemed commercially successful, they might neglect or even suppress artistic elements that don't align with these criteria. This could further narrow the scope of what is considered "acceptable" art, potentially stifling innovation and limiting the cultural dialogue that art is meant to inspire.

In the landscape of art and technology, the integration of AI into the creative process marks a transformative moment, echoing shifts that have characterized the relationship between humans and machines. From the rudimentary dialogues of ELIZA to the sophisticated capabilities of generative AI models such as ChatGPT, we have observed a remarkable journey of technological innovation and the complex interplay between human creativity and machine intelligence.

As we have seen, the blurring boundaries between AI-generated content and human-created outputs raise ethical and philosophical dilemmas. The capability that allows generative AI to produce novel outcomes can also be its Achilles's heel. Without careful oversight, generative systems can produce misleading and sometimes harmful outputs. Navigating the landscape of generative AI requires a judicious blend of enthusiasm for its capabilities and caution against its pitfalls.

The promise of AI in art is immense, offering new avenues for creative expression and collaboration. However, this promise comes with economic, ethical, and creative challenges. The risk of displacement of human artists, the ethical quandaries surrounding data use, and the potential for a loss of creative diversity are issues that cannot be ignored.

In essence, on the one hand, we stand at the cusp of a new creative renaissance, fueled by the symbiotic relationship between humans and machines. On the other hand, while generative AI offers exciting possibilities for art creation, it also brings forth a set of challenges and controversies that the art world, and society at large, must grapple with. As we as a society continue to explore this intersection of technology and creativity, these debates will shape the future trajectory of AI in art. While AI has the potential to be a valuable tool in the artistic process, there's a growing concern that its influence could inadvertently lead to a more homogenized and less adventurous artistic landscape. It's precisely in navigating these complexities that we find the most compelling opportunities for human-AI collaboration. By approaching AI as a tool that can augment rather than replace human creativity, we can harness its capabilities to enrich the artistic process, broaden cultural narratives, and explore new frontiers in artistic expression. As we stand at this intersection of art, technology, and society, the choices we make today will shape the future of creative endeavors.

SCIENTIFIC RESEARCH AND INNOVATION

We live in a period of time when technological leaps are molding our reality into scenes that were reserved for science fiction. The paradigms of learning, teaching, and scholarly investigation are changing, with AI serving as both the instigator and the facilitator of this evolution in research and innovation (Crawford & Calo, 2016; Furman & Seamans, 2019).

The exponential growth of information and the rapid emergence of new fields of study have surpassed human capacity to keep up (Stevens, 2015). Educational systems, erected on historical models, face the formidable challenge of preparing students for a future job market that is speculative in nature and is reliant on nascent technologies. Similarly, the complexity of modern research, often requiring an interdisciplinary approach, calls for resources and computational capabilities that are frequently beyond what individual scholars or institutions can muster.

In this chapter, we will explore the transformative role of artificial intelligence in augmenting human capabilities in business and education. We will focus on how generative AI can act as an ally in accelerating research efforts and driving innovation. We will also examine the implications and challenges that arise as we usher in the era of human–AI collaboration.

IMPACT IN RESEARCH AND INNOVATION

As we face the Fourth Industrial Revolution, marked by rapid advancements in artificial intelligence, robotics, and other emerging technologies, the following question arises: can machines be our new collaborators in pushing the boundaries of scientific research and innovation?

The answer is a resounding "yes". Collaborative AI systems designed to work in tandem with humans are capable of accelerating scientific discoveries and driving innovation. This includes a synergistic relationship in which both humans and AI contribute their unique strengths to solve complex problems (Biswas et al., 2001).

Think of it as a high-stakes brainstorming session where your co-contributor reads millions of academic papers in seconds, identifies patterns you cannot see, and never gets tired.

High-level research has traditionally been the domain of well-funded institutions and individuals with specialized training. Collaborative AI has the potential to democratize this landscape. With intuitive interfaces and the ability to handle complex calculations, these systems can make advanced research tools accessible, leveling the playing field and fostering a more inclusive environment for innovation.

Time and resources are often the most significant barriers to scientific progress. Traditional research methods can be slow and laborious, requiring years or even decades to yield actionable insights. Traditional research is a multi-step endeavor that has been fine-tuned over centuries. It often starts with the identification of a research gap or problem, followed by an exhaustive review of existing literature to understand what is already known about the subject. Researchers then formulate hypotheses and design experiments or studies to test these hypotheses, a step that requires meticulous planning and a keen understanding of scientific methods.

The execution of the research plan is a labor-intensive process that involves data collection, which could range from conducting surveys and interviews to running lab experiments or field studies. Once the data is collected, researchers spend considerable time analyzing it, often using statistical models to interpret the results. The culmination of this rigorous process is the research publication, which undergoes peer review to ensure its validity and contribution to the field.

Research is rarely a solo endeavor; it is often carried out by teams comprising individuals with diverse skill sets. Team members contribute different perspectives and expertise, enabling a more holistic approach to problem-solving. Research, particularly in scientific and technical fields, often requires substantial financial investment for equipment, data collection, and manpower. Funding, whether from governmental agencies, private institutions, or corporate entities, is the lifeblood that sustains the research ecosystem.

Now, let's pivot to how AI can revolutionize this traditional model. While the basic steps of research might remain constant, AI can significantly accelerate each phase and add new dimensions to the research process. With previously trained AI plugins, researchers can use AI to conduct a literature review in a fraction of the time it would take a human, using natural language processing to summarize key findings and identify gaps. In the experimental design stage, AI can simulate various scenarios, helping researchers optimize their methods for cost, time, or accuracy. AI can significantly enhance R&D operations by assisting in idea generation and automating data analysis, thereby accelerating the discovery process.

When it comes to execution, AI can automate data collection processes, especially in fields of social sciences, biology, and healthcare, where large datasets are common (Steels & Brooks, 2018; Damiano & Stano, 2023; Elliott, 2021). For analysis, machine learning algorithms can identify complex patterns and relationships in the data that might be impossible for a human to discern. These capabilities make AI a collaborator capable of contributing ideas and suggesting alternative hypotheses or interpretations.

The implications for team dynamics can also be quite significant. With AI handling some of the more tedious and time-consuming tasks, human team members can focus on creative and complex aspects of research. This could lead to smaller, more agile research teams and make funding more efficient, as AI can perform many tasks more quickly and accurately, reducing the overall cost and time required for research projects. At the same time, the utilization of AI tools can streamline the management of larger teams by automating various administrative tasks and facilitating real-time, data-driven decision-making, ultimately making team management more effective and responsive.

For a solo researcher, the advent of AI technologies can be a game-changer, leveling the playing field in ways that were once the exclusive domain of well-funded research teams. By assuming the role of a multi-skilled research assistant, AI can handle a range of tasks that traditionally would have required a team of specialists. From scouring academic journals for relevant literature to running complex statistical analyses, AI can manage various stages of the research process, effectively acting as a one-stop research partner. The integration of AI into the research process thus can be a powerful augmentation of human capabilities. It may make research faster, cheaper, and possibly more innovative, allowing us to tackle complex problems with an amplified level of sophistication and efficiency.

For the modern business professional, artificial intelligence offers an array of tools that can significantly elevate the quality and impact of their work. For example, a recent study conducted by Boston Consulting Group suggests that groups that used GPT-4 while working on a variety of tasks performed 40% better than groups that didn't (Martines, 2023).

Generative AI can assist in numerous ways. It can conduct a rapid analysis of the latest industry trends, competitor strategies, and market data, providing a well-rounded view of the landscape. Let's consider using AI to assist in writing a report. Utilizing natural language generation, AI can draft a preliminary version of the report, organize it around key themes, and enrich it with relevant data points. AI can suggest rhetorical devices and storytelling elements, and even add some humor to make the report more engaging and memorable. The owner of the report can then refine the draft, adding their personal touch and expertise.

AI can perform tasks ranging from data visualization to presentation design. Plugins now allow inputting raw data and receiving back a series of professionally designed charts and graphs that display the information but also highlight key trends and insights (Triantoro, 2023). AI can suggest the optimal sequence for your presentation slides based on storytelling structures, ensuring that the presentation has a logical flow and maximum impact for the selected type of audience.

Beyond reports and presentations, AI can assist in other business tasks. For instance, if you are pitching to investors, AI can analyze past successful pitch decks and investor behavior to suggest what content to include and emphasize. It can generate financial models based on your inputs, offering projections and suggesting strategies for revenue growth or cost reduction. It can scan social media and customer reviews to provide real-time insights into brand perception, invaluable for any marketing presentation. In team settings, AI can act as a project manager, tracking progress, flagging delays, and suggesting reallocations of resources to ensure deadlines are met.

Thus, AI can dramatically enhance a business person's effectiveness by handling time-consuming tasks and offering intelligent suggestions for improvement. In addition to augmenting the individual skill sets of business professionals, AI can serve as a powerful catalyst for organizational innovation. AI can analyze market trends, consumer behavior, and emerging technologies to suggest new product features or entirely new lines of business. It can simulate a variety of scenarios to predict the success of a new venture, reducing the risk associated with innovation. Furthermore, AI can facilitate innovation sprints, to rapidly prototype and test multiple ideas, allowing companies to quickly identify the most promising avenues for development. In this way, AI extends the innovative capacity of businesses, allowing proactive disruption, capable of setting new standards and shaping consumer expectations. Essentially, AI becomes a co-creator, driving innovation with a speed and precision that sets new benchmarks for competitive advantage.

IMPACT IN EDUCATION

While the transformative impact of artificial intelligence in research and innovation is indisputable, it is important to recognize where the seeds of these transformations are sown—the area of education. Education is the bedrock to build advancements in research and innovation, and it is the place to cultivate the essential skills of students. Understanding how AI can democratize and enrich the educational experience offers a glimpse into the future of learning and provides valuable insights into how upcoming generations will interact with AI in research and innovation contexts.

AI is becoming a transformative force with the potential to democratize learning and redefine the boundaries of pedagogical practice (Ng et al. 2022). For centuries, the educational system has been a one-size-fits-all approach, standardized curricula, and a limited scope for individualized learning. Yet, in the age of AI, we see the trends that could alter the way education is conceived, delivered, and experienced.

Democratization is one of the most compelling aspects of integrating AI into education. Historically, quality education has often been a privilege of the few, limited by geographical location, socio-economic status, and access to skilled educators. AI can shatter these barriers. With platforms that can offer personalized learning experiences, a child in a remote village can have access to the same quality of education as a student in a bustling metropolis. Adaptive learning algorithms can tailor educational content to suit individual learning styles and paces, offering a customized education that is both engaging and effective. For adult learners, AI-powered platforms can provide on-demand courses, reskilling programs, and career advice, making lifelong learning a realistic and accessible goal.

AI revolutionizes the act of learning itself. Intelligent tutoring systems could assess not only what you know but also how you think. These systems can pose complex problems, encourage open-ended discussions, and offer real-world scenarios for problem-solving, fostering a deeper, more analytical mode of learning.

However, the integration of AI into education has its challenges. As some of us marvel at the capabilities of generative AI, an important question emerges: "Wow, 'we've got generative AI – should we now stop thinking?" It is a complex question that encapsulates a genuine concern about the potential for AI to not just augment but also modify human cognitive processes.

With AI tools capable of crafting persuasive essays, solving complex mathematical problems, and generating research hypotheses, there is a temptation to lean too heavily on these technologies. Could we reach a point where students, enamored by the efficiency and accuracy of AI, become passive recipients rather than active participants in their educational journey? This can undermine the development of critical thinking and create a generation that lacks the ability to question, challenge, and innovate independently of algorithmic aid.

Let's wind the clock back a few decades to when calculators were first introduced into classrooms. The uproar was significant. Educators protested, arguing that these gadgets would erode fundamental arithmetic skills, render students lazy, and undermine the purpose of mathematics education (Hembree & Dessart, 1986; Trouche, 2005). Fast-forward to today and calculators are not just tolerated but also embraced as standard educational tools. They are seen as aids that allow students to focus on more complex problem-solving and conceptual understanding.

This trajectory of skepticism to acceptance is not unique to calculators. We have seen this pattern repeated as technology progressively integrates with various domains. It is evident in the fields of machine learning and data analytics, where computational tools are considered the gold standard (So et al. 2020). Sophisticated algorithms sift through massive datasets, perform complex analyses, and generate predictive models with a level of accuracy and efficiency that would be humanly impossible or, at the very least, incredibly time-consuming.

Interestingly, this trust in machines extends beyond acceptance; we now find ourselves trusting machines more with numbers than we trust ourselves. Why is this the case? For one, the sheer computational power of machines allows them to handle large datasets and complex equations far more quickly and accurately than a human could. The iterative and consistent nature of machine processing eradicates the chances of human errors which might result from fatigue, cognitive biases, or simple miscalculations (Jordan & Mitchell, 2015).

New York City schools banned ChatGPT technology in schools in early 2023, just to bring it back at the end of the school year (Rosenblatt, 2023). It seems history repeats itself. But how could modern generative AI systems and LLMs be different from calculators several decades back? The initial ban and subsequent reinstatement of ChatGPT technology in New York City schools reminds us that every technological advance brings with it a period of uncertainty, resistance, and ultimately, adaptation. But how do modern generative AI systems and Large Language Models differ from the calculators that faced similar scrutiny decades ago?

The capabilities of generative AI extend far beyond the computational utility of calculators. While calculators are designed for specific tasks, primarily arithmetic and basic mathematical functions, generative AI is multi-faceted tools with a broad spectrum of applications. They can draft essays, summarize complex articles, generate research questions, and engage in nuanced conversations. In essence, their utility is not confined to a single subject or skill but permeates various facets of education, from language arts to social studies and beyond.

Another point of distinction is in the level of personalization. Calculators are static tools; they perform the same functions regardless of who is using them. LLMs, on the other hand, have the potential for adaptive learning. They can tailor their interactions based on the user's needs, skill level, and learning style. Over time, they can become more effective educational aids, offering personalized feedback and resources that can significantly enrich the learning experience.

However, the most striking difference between calculators and Large Language Models is in the latter's ability to converse using words. For humans, numbers are often abstract entities; they are vital for quantification and analysis but are not the primary medium through which we experience or interpret the world. It is relatively straightforward to outsource numerical tasks to calculators because numbers, in many contexts, lack the emotional and cognitive texture that words possess. We can entrust machines with numbers because doing so does not impinge upon our understanding of ourselves or the world around us in any fundamentally human way.

Words, however, are a different matter entirely. Humans think in narratives, symbols, and meanings—all of which are constructed using words. Language is not just a tool for communication; it is the fabric of our consciousness, the framework within which we make sense of our existence. Words are magical to us, laden with history, culture, and personal experiences that give them deep, often ineffable, meanings. When machines begin to speak, to converse in words that carry the weight of human experience, they cross a threshold that is unsettling and exhilarating at the same time.

This capacity for linguistic interaction triggers anthropomorphic feelings in us; we begin to imbue these machines with human-like qualities, such as intelligence or even the capacity for understanding and empathy (Salles et al. 2020; Troshani et al. 2021). However, anthropomorphism can have both beneficial and adverse outcomes. On the one hand, it makes interactions with machines more natural and engaging, potentially enriching our educational experiences and making technology more accessible. On the other hand, it can lead to misplaced trust or ethical ambiguity. Is it appropriate, for example, for students to consult a generative AI tool for personal or ethical advice? If a machine can generate a compelling narrative, what happens to the value we place on human storytelling or creative writing?

Another critical aspect is how much trust can we, or should we, assign to AI? This is not a trivial question, especially given AI's propensity to hallucinate, or generate outputs that are factually incorrect or misleading. While it is tempting to view AI systems as oracles of wisdom and knowledge, their limitations are a sobering reminder that they are far from infallible.

The notion of trust in AI is a complex mix of accuracy, reliability, ethics, and emotional resonance. On the surface, AI models can appear remarkably accurate, generating text that is grammatically correct and contextually relevant. They can summarize research papers, answer factual questions, and engage in debates. However, this exterior of competence can sometimes mask an interior that is less reliable. AI models, for all their sophistication, can generate outputs that are factually incorrect, contextually misleading, or ethically problematic. These hallucinations are symptomatic of the AI's lack of understanding and inability to comprehend the nuance and depth of human experience and knowledge (Schank, 1987; Chivers, 2019).

This propensity for error may have significant implications for trust. For example, in a classroom setting, educators will have to verify the accuracy of an AI-generated summary of a historical event or a scientific theory. In a research context, scholars who rely on LLMs to generate hypotheses or interpret data will have to employ a significant amount of human oversight. While AI can be an incredibly powerful tool for augmenting human capabilities, it cannot yet replace the depth of understanding, the ethical reasoning, and the critical thinking that define human intelligence. Trust in AI, therefore, must be conditional and contextual. It should be seen not as a binary state, but as a spectrum that varies based on the task at hand, the stakes involved, and the limitations of the specific AI model being used.

Critically assessing trust in AI also involves acknowledging the dynamic nature of both technology and human understanding. AI is continually evolving, with newer models being trained on more extensive datasets and designed to minimize biases and errors. Similarly, our understanding of AI, its capabilities, and its limitations is also deepening. Trust is a dynamic relationship that evolves, and our trust in AI will change as our experience with AI deepens.

As we think about the role of AI in education, students, educators, and parents confront a question that is practical and timely: "What do I, as a student, actually learn if AI writes for me?" This question opens a Pandora's box of concerns that echo far beyond the classroom and reverberate through our broader understanding of education, skill development, and human cognition. The most immediate concern is: will we know how to write? If AI can draft essays, summarize articles, and generate research papers, is there a risk that students will become just spectators in their educational journey, outsourcing the hard work of thinking, synthesizing, and articulating to algorithms? The parallel with calculators is instructive but not entirely comforting. While calculators did not eliminate the need for basic arithmetic skills, they did fundamentally alter the landscape of mathematics education. Advanced calculators can perform complex functions, from calculus to statistical analysis, raising the question: how many people today, unaided by technology, can solve complex math problems? Similarly, if AI takes over the act of writing, the skill could become another casualty in the growing list of abilities that technology has made obsolete.

But the concerns do not stop at technical skills; they extend into cognitive and emotional development. Writing is not only a mechanical act of putting words on paper, but it is a deeply cognitive process that involves organizing thoughts, constructing arguments, and expressing emotions. Will we know how to express our thoughts without first consulting with AI? If the answer is no, or even maybe, then we face a crisis that goes beyond education and touches on the essence of human agency. Relying on AI to articulate our thoughts could lead to a form of cognitive outsourcing, where we risk losing the ability to reflect, reason, and express ourselves independently.

The implications of this could be deep and not only for individuals but for society at large. The ability to think critically and express oneself clearly is not just an academic skill. In a world increasingly dominated by complex issues, from climate change to social justice, the ability to articulate thoughtful opinions is crucial. If we outsource this skill to AI, do we also risk outsourcing our civic responsibility, our social consciousness, and even our humanity?

IN AI WE TRUST

As AI technologies make their way into professional life, those working their way up the career ladder are faced with a perplexing conundrum: how much can one trust the output of the advanced AI systems? This question is becoming a universal concern that spans industries and job functions. From marketing executives using AI to optimize campaigns to lawyers employing machine learning algorithms for case research, the implications of trusting AI output have far-reaching professional and ethical ramifications.

The issue of trust becomes especially thorny when one considers the phenomenon of AI hallucinations, the instances when the system generates outputs that are factually incorrect, logically flawed, or ethically dubious. While some may argue that these hallucinations are similar to human errors and can be corrected over time, it is important to recognize that the stakes are often high in professional settings. A hallucination in a financial model could lead to disastrous investment decisions, just as a factual error in a legal brief could compromise a case (Milmo, 2023). Given these risks, the uncritical acceptance of AI output could be professionally irresponsible, if not downright dangerous.

The issue of hallucinations brings up an interesting point: not all hallucinations are bad, especially when we are looking for imaginative or creative outcomes. If you are in a creative field such as advertising, design, or strategic planning, an AI that thinks outside the box could offer innovative solutions that a human might not conceive. In these scenarios, the machine's ability to deviate from the norm might actually be an asset rather than a liability. However, there are advantages and disadvantages. While AI hallucinations can be useful for brainstorming sessions, they are not always optimal for concrete outcomes that require precision, reliability, and factual accuracy.

The dichotomy between imagination and concreteness poses a challenge: how do we harness the creative potential of AI while also mitigating the risks associated with its propensity for error? The answer lies in a balanced, critical approach to AI adoption. Professionals must be trained not just to use AI tools but also to understand their limitations. Processes should be put in place to double-check AI outputs against human expertise and ethical standards. Essentially, the relationship between professionals and AI should be one of critical collaboration rather than blind reliance.

It is important to point out that the learning mechanism of generative AI tools is a process of statistical approximation rather than genuine understanding. These models are trained on massive datasets composed of text from various sources, such as books, articles, websites, and social media posts. The training process involves adjusting millions or even billions of parameters to predict the next word in a sequence based on the words that precede it (Cybellium Ltd, 2023; Anderson & Coveyduc, 2020). However, this form of learning lacks comprehension. The machine doesn't understand language, but instead it calculates probabilities. It has no sense of context beyond the dataset it was trained on, and it does not grasp the ethical, cultural, or social nuances that permeate human communication.

The source from which AI models learn human language brings its own set of challenges. Because the training data often comes from the internet, it is a mixed bag of factual information, opinions, biases, and sometimes, outright misinformation (Paschen et al. 2020). This raises concerns about the reliability and ethical integrity of AI-generated content. Can a tool trained on such a varied dataset be considered a trustworthy source of information or advice?

This brings us to the notion of the "source of truth" in the digital age. Traditionally, encyclopedias such as Britannica served as trusted repositories of curated knowledge. With the advent of the Internet, Wikipedia emerged as a more dynamic, but less controlled, source of information (Reagle & Koerner, 2020). Now, tools like ChatGPT are entering the scene, capable of generating information on-the-fly. Each transition represents a shift not just in the medium but also in the epistemological foundations of how we define "truth." While Britannica relied on expert curation, Wikipedia introduced the concept of crowd-sourced wisdom, and ChatGPT brings algorithmic generation into the mix.

In an era where traditional gatekeepers of information are being bypassed or augmented by AI, the lines between expert opinion, crowd wisdom, and algorithmic output are increasingly blurred. This democratization of information has its merits, offering more people access to a broader array of perspectives. However, it also complicates the task of discerning the quality of information. When everyone has a voice, and algorithms can emulate those voices, how do we decide what to trust?

Critically, as we rely more on AI tools like ChatGPT for information and even decision-making, there is a risk of cognitive outsourcing. Will we become less critical consumers of information, trusting the algorithm to do the thinking for us? Moreover, the algorithm's inherent biases and limitations could propagate through our decision-making processes, leading to outcomes that are skewed or ethically questionable.

The proliferation of AI capabilities in generating creative works, be it text, music, or visual art, introduces a complex terrain of legal, ethical, and philosophical quandaries. For example, the current U.S. copyright framework is rooted in a human-centric notion of creativity, as outlined in the Guidance for Works Containing Material Generated by Artificial Intelligence (U.S. Copyright Office, 2023). The term "author" explicitly refers to a human being, thereby excluding non-human entities such as AI systems. This legal stance is a reflection of longstanding cultural and philosophical views that place human creativity on a unique pedestal, inherently worthy of legal protection.

However, the boundaries between human and machine-generated creations are getting increasingly porous. In many instances, what we encounter is not purely machine-generated work but rather a hybrid, a result of human–AI collaboration. The prompt, the guiding principle, or even the fine-tuning of the AI system often comes from the human mind. In such cases, can we not argue that the final product is imbued with human creativity, thereby meriting copyright protection?

This is a compelling argument for several reasons. First, it recognizes that creativity is not a binary attribute that you either have or do not have. Creative process is a spectrum. Even in traditional artistic processes, tools and external influences play a role. A painter uses a brush, a writer uses a keyboard, and these are tools, much like an AI, that facilitate the creative process. Second, the human-generated prompt or guidance often sets the creative direction, which means that the human is still the "author" in a conceptual sense, even if the AI performs the mechanical aspects of the creation.

CONVERGING MINDS. A NEW PARADIGM

The convergence of human and machine intelligence offers a novel paradigm for addressing complex problems, from scientific research to ethical dilemmas. This paradigm, which in this book we call "Converging Minds," involves the creation of hybrid cognitive models that integrate human intuition with machine analysis. It is the future where researchers, policymakers, and artists could tap into the hybrid models to explore various scenarios, generate hypotheses, and create art. These models could serve as advanced thought laboratories, providing a space for human–machine teams to collaboratively explore the boundaries of what is possible, ethical, and beautiful.

The concept of Converging Minds serves as a turning-point moment in our evolving relationship with technology. This paradigm shift moves us from a dualistic framework, where human and machine intelligences operate in parallel but separate domains, to a more integrated, synergistic model. But what does this convergence mean, and more importantly, what are the implications of creating hybrid cognitive models that meld human intuition with machine analysis?

In the scientific domain, the benefits of such a paradigm are ambitious. Imagine a scenario where a human researcher formulates a complex question, related to climate change or medical diagnostics. Using the Converging Minds framework, the researcher could input this question into a hybrid cognitive model, which would then generate a range of hypotheses, backed by data and probabilistic reasoning. The researcher could then refine these hypotheses based on human intuition, ethical considerations, or the current scientific consensus, factors that the machine might not fully grasp. The result is a truly collaborative research methodology that leverages the strengths of both human and machine cognition.

In the area of art and creativity, the Converging Minds paradigm opens up new avenues for exploration. Artists could collaborate with AI to create new works. AI could generate the outline of a novel or the melody of a song based on certain parameters set by the artist. The human artist could then infuse these outputs with emotional depth, nuance, and cultural relevance, transforming them into something profoundly human yet technologically advanced.

At the same time, the ethical implications of Converging Minds are equally complex: machines can introduce a variety of biases into decision-making processes and pose existential questions to human creativity. Thus, the final decision should always be made by humans, who can consider factors such as empathy, social justice, and moral responsibility, which are currently beyond the machine's understanding.

In sum, the Converging Minds paradigm offers an approach to solving complex problems, fostering creativity, and making ethical decisions. This convergence is a tool that amplifies our capabilities while also magnifying our responsibilities. It challenges us to redefine the boundaries of intelligence, creativity, and ethics, forcing us to confront the limitations and biases inherent in both human and machine cognition. As we navigate this uncharted territory, our success will depend not only on technological innovation but also on our ability to integrate these advances with the wisdom, ethics, and critical thinking that define us as humans.

REFERENCES

Abdulkareem, M., & Petersen, S. E. (2021). The promise of AI in detection, diagnosis, and epidemiology for combating COVID-19: Beyond the hype. *Frontiers in Artificial Intelligence*, *4*(May), 652669.

Anderson, J. L., & Coveyduc, J. L. (2020). *Artificial Intelligence for Business: A Roadmap for Getting Started with AI*. John Wiley & Sons.

Biswas, G., Schwartz, D., Bransford, J., & Vanderbilt, Teachable Agents Group at. (2001). Technology support for complex problem solving: From SAD environments to AI. In K. D. Forbus & P. J. Feltovich (Eds.), *Smart Machines in Education: The Coming Revolution in Educational Technology* (483pp, pp. 71–97, vi). The MIT Press. https://dl.acm.org/doi/abs/10.1145/3641289

Bryson, J. J., Diamantis, M. E., & Grant, T. D. (2017). Of, for, and by the people: The legal lacuna of synthetic persons. *Artificial Intelligence and Law*, *25*(3), 273–291.

Carter, S., & Nielsen, M. (2017). Using artificial intelligence to augment human intelligence. *Distill*, *2*(12). https://doi.org/10.23915/distill.00009.

Chivers, T. (2019). *The AI Does Not Hate You: The Rationalists and Their Quest to Save the World*. Weidenfeld & Nicolson.

Cohn, J. (2021). 'The Scientist of the Holy Ghost': Sunspring and reading nonsense. *CJEM*, *60*(5), 1–21.

Crawford, K., & Calo, R. (2016). *There Is a Blind Spot in AI Research*. Nature Publishing Group UK, 13 October. https://doi.org/10.1038/538311a.

Cybellium Ltd. (2023). *Mastering AI Model Training*. Cybellium Ltd.

Damiano, L., & Stano, P. (2023). Explorative synthetic biology in AI: Criteria of relevance and a taxonomy for synthetic models of living and cognitive processes. *Artificial Life*, *29*(3), 367–387.

Du Sautoy, M. (2020). *The Creativity Code: Art and Innovation in the Age of AI*. Harvard University Press.

Elliott, A. (2021). *The Routledge Social Science Handbook of AI*. Routledge.

Fink, T., & Akdag Salah, A. A. (2023). Extending the visual arts experience: Sonifying paintings with AI. In C. Johnson, N. Rodríguez-Fernández, & S. M. Rebelo (Eds.), *Artificial Intelligence in Music, Sound, Art and Design* (pp. 100–116). Springer Nature Switzerland.

Furman, J., & Seamans, R. (2019). AI and the economy. *Innovation Policy and the Economy*, *19*(January), 161–191.

Goenaga, M. A. (2020). A critique of contemporary artificial intelligence art: Who is Edmond de Belamy? *AusArt*, *8*(1), 49–64.

Hageback, N. (2021). *AI for Creativity*. CRC Press.

Harris, J. D., & Waggoner, B. (2019). Decentralized and collaborative AI on blockchain. In *2019 IEEE International Conference on Blockchain (Blockchain)* (pp. 368–375). IEEE. https://arxiv.org/abs/1907.07247

Hembree, R., & Dessart, D. J. (1986). Effects of hand-held calculators in precollege mathematics education: A meta-analysis. *Journal for Research in Mathematics Education*, *17*(2), 83–99.

Jordan, M. I., & Mitchell, T. M. (2015). Machine learning: Trends, perspectives, and prospects. *Science*, *349*(6245), 255–260.

Luce, L. (2019). Democratization and impacts of AI. In *Artificial Intelligence for Fashion* (pp. 185–195). Apress.

Luchs, I. (2023). AI for all? Challenging the democratization of machine learning. *A Peer-Reviewed Journal About*, *12*(1), 135–147.

Martines, D. (2023). Decoding the jagged frontier: AI, centaurs, and cyborgs, and the future of work. LinkedIn, 19 September.

Milmo, D. (2023). Two US lawyers fined for submitting fake court citations generated by ChatGPT. *The Guardian*, 23 June.

Monaco, N., & Woolley, S. (2022). *Bots*. John Wiley & Sons.

Ng, D. T. K., Luo, W., Chan, H. M. Y., & Chu, S. K. W. (2022). Using digital story writing as a pedagogy to develop AI literacy among primary students. *Computers and Education: Artificial Intelligence*, *3*(January), 100054.

Paschen, U., Pitt, C., & Kietzmann, J. (2020). Artificial Intelligence: Building blocks and an innovation typology. *Business Horizons*, *63*(2), 147–155.

Reagle, J., & Koerner, J. (2020). *Wikipedia@ 20: Stories of an Incomplete Revolution*. The MIT Press.

Riedl, M., Thue, D., & Bulitko, V. (2011). Game AI as storytelling. In P. Antonio González-Calero & M. Antonio Gómez-Martín (Eds.), *Artificial Intelligence for Computer Games* (pp. 125–150). Springer New York.

Rosenblatt, K. (2023). ChatGPT ban dropped by New York City Public Schools. *NBC News*, 18 May.

Salles, A., Evers, K., & Farisco, M. (2020). Anthropomorphism in AI. *AJOB Neuroscience*, *11*(2), 88–95.

Schank, R. C. (1987). What is AI, anyway? *AI Magazine*, *8*(4), 59–59.

Seok, B. (2023). Chapter five the uncharted world of AI art: Music and AI Bongrae Seok. In *Venturing into the Uncharted World of Aesthetics*, 68.

So, A., Joseph, T. V., Thas John, R., Worsley, A., & Asare, S. (2020). *The Data Science Workshop: A New, Interactive Approach to Learning Data Science*. Packt Publishing Ltd.

Steels, L., & Brooks, R. (2018). *The Artificial Life Route to Artificial Intelligence: Building Embodied, Situated Agents*. Routledge.

Stephensen, J. L. (2019). Towards a philosophy of post-creative practices? – Reading Obvious' 'Portrait of Edmond de Belamy.' *Politics of the Machine Beirut 2019*, *2*, 21–30.

Stevens, C. (2015). *On the Acceleration of Human Evolution and Other Writings*. CreateSpace Independent Publishing Platform.

Triantoro, T. (2023). ChatGPT gets chatty with data: How generative AI models. LinkedIn, 31 August.

Troshani, I., Hill, S. R., Sherman, C., & Arthur, D. (2021). Do we trust in AI? Role of anthropomorphism and intelligence. *Journal of Computer Information Systems*, *61*(5), 481–491.

Trouche, L. (2005). Calculators in mathematics education: A rapid evolution of tools, with differential effects. In D. Guin, K. Ruthven, & L. Trouche (Eds.), *The Didactical Challenge of Symbolic Calculators: Turning a Computational Device into a Mathematical Instrument* (pp. 9–39). Springer US.

Tsigelny, I. F. (2019). Artificial intelligence in drug combination therapy. *Briefings in Bioinformatics*, *20*(4), 1434–1448.

U.S. Copyright Office. (2023). Copyright registration guidance: Works containing material generated by Artificial Intelligence. https://copyright.gov/ai/ai_policy_guidance.pdf

Weizenbaum, J. (1966). ELIZA – A computer program for the study of natural language communication between man and machine. *Communications of the ACM*, *9*(1), 36–45.

Weizenbaum, J., & McCarthy, J. (1977). Computer power and human reason: From judgment to calculation. *AIP*. https://www.amazon.com/Computer-Power-Human-Reason-Calculation/dp/0716704633

Yu, Z., Cen, G., Zhao, L., & Zhu, C. (2023). Design of smart mistake notebook based on AI and big data. In M. M. Rodrigo, N. Matsuda, A. I. Cristea, & V. Dimitrova (Eds.), *Computer Science and Education* (pp. 18–28). Springer Nature Singapore.

2 The Core and Ecosystem of Generative AI

In the introductory chapter, we explored the synergy between human intelligence and artificial mechanisms, underlining the transformative potential of their collaboration. In this chapter, we will take a deeper look into advanced machine learning (ML) models that serve the purpose of generating new content and potentially have a very collaborative nature. We will explore the foundational concepts of supervised and unsupervised learning, which serve as the bedrock for generative models like generative adversarial networks (GANs) and variational autoencoders (VAEs). While these generative models primarily operate within the area of unsupervised learning, elements of supervised learning can also be incorporated to enhance their performance. We will thus also look into how Reinforcement Learning (RL), another distinct methodology, can be synergistically used to optimize the training processes of GANs and VAEs.

You may ask, "Where does ChatGPT fit in?" We will also elaborate on transformer algorithms that became the foundation of large language models (LLMs) exactly like Generative Pre-trained Transformer (GPT) (Babcock & Bali, 2021; Bond-Taylor et al., 2022; Salehinejad et al., 2017; Fröhling & Zubiaga, 2021). Generative AI and LLMs like ChatGPT are closely related but not the same. Both are designed to generate new content and use ML algorithms for this purpose. However, generative AI is a broader category that includes various types of models, such as those for generating images or music. LLMs are a specific type of generative AI focused on text. The architecture and training methods can differ between the two, as can the type of data they are trained on.

In order to understand LLMs better, we will introduce an example of one of the most important algorithms to build them: a transformer. Not all language models rely on transformers. While transformers have become popular for their effectiveness in handling a wide range of language processing tasks, there are other architectures as well. Earlier models like recurrent neural networks (RNNs) and long short-term memory (LSTM) networks that we will cover too have also been used for language modeling. But transformers are those algorithms that have recently gained most prominence due to their scalability and ability to handle long-range dependencies in text, but they are not the only option for building language models.

To avoid overwhelming the reader, we will clearly delineate the scope and applicability of each methodology and algorithm, highlighting their unique contributions as well as their interconnections. By the end of this chapter, the readers will hopefully gain a comprehensive understanding of how these diverse AI methodologies coalesce to advance the field of generative AI.

Why are the technicalities necessary? Well, understanding the technology and models behind advanced and generative AI is analogous to knowing the ingredients

40

DOI: 10.1201/9781032656618-3

and methods in cooking; while you can enjoy a meal without this knowledge, understanding the processes can enrich the experience and prevent potential pitfalls. Imagine a video making rounds on the internet where a renowned world leader makes controversial statements that could have global political implications. The video looks incredibly realistic, the voice matches, and the nuances in expressions are impeccable. But there's a catch – it is not real. It is a product of "deepfake" technology, powered by advanced generative AI models like GANs. For the average person who is unaware of the capabilities of modern AI, this video might be taken at face value, leading to potential misunderstandings, spread of misinformation, and political tensions. However, someone versed in the technology behind the scenes would approach such content with a healthy dose of skepticism. They would understand that AI models today can synthesize hyperrealistic content, from images to videos to voices (Firc et al., 2023; Ahmed & Chua, 2023; Meikle, 2022).

What is more, beyond mere awareness, understanding the models also means knowing their limitations. For instance, while deepfakes can be eerily accurate, they are not perfect. Certain glitches, inconsistencies, or artifacts can give them away, but only if one knows what to look for. By understanding the technology, one can better discern facts from fiction, reducing the potential harm of such content. On the flip side, for creators and technologists, understanding the models opens a world of potential. The same technology that can be misused for misinformation can be harnessed for benign and innovative applications, such as in films, art, and historical recreations.

ML LANDSCAPE

Let us begin then. Before we dwell into GANs, VAEs, and LLMs, we need to take a step back to the basics. All these complex algorithms belong to a broader algorithmic family called machine learning. Machine learning (ML) is a field of computer science that gives computers the ability to learn without being explicitly programmed (El Naqa & Murphy, 2015; Alpaydin, 2021). ML algorithms are trained on data, and then they can use that data to make predictions or decisions. There are two basic types of ML philosophies: supervised learning and unsupervised learning.

- **Supervised learning** is a type of ML where the algorithm is given labeled data. This means that the data has been tagged with the correct output. For example, if you are training an algorithm to classify images of cats and dogs, the data would be labeled with either "cat" or "dog" (Cunningham et al., 2008; Nasteski, 2017).
- **Unsupervised learning** is a type of ML where the algorithm is not given labeled data. This means that the algorithm has to find patterns in the data on its own. For example, if you are training an algorithm to cluster images, the algorithm would have to find groups of images that are similar to each other by itself, without human intervention (Ngiam et al., 2011; *Unsupervised Learning Algorithms*, n.d.).
- **Deep learning** is a type of unsupervised learning technique that uses artificial neural networks (ANNs) to learn from data. ANNs are inspired

by the human brain, and they are able to learn complex patterns in data (Hopfield, 1988; Abraham, 2005; da Silva et al., 2016). At its core, deep learning involves training multilayer neural networks on large amounts of data. These layers, often dozens or hundreds deep, allow the model to learn from raw data, abstracting increasingly complex features at each layer. This hierarchical learning capability enables deep learning models to excel in tasks that require understanding structures, in language, images, and even abstract concepts. The strength of deep learning is in its predictive power and in its ability to learn from unstructured data, thereby removing the need for manual feature extraction, which is a limitation of traditional ML techniques. Deep learning's applications are vast, ranging from natural language understanding to computer vision, and one of its most fascinating branches is generative AI.

• **RL** is a type of ML paradigm where an agent learns how to behave in an environment by performing actions and observing the rewards of those actions. The goal is to find a policy that will allow the agent to take actions in states of the world in a way that maximizes some notion of cumulative reward. It's widely used in various applications like game playing, robotics, and natural language processing (NLP), and even in stock trading, whereas deep learning is primarily concerned with algorithms and models for learning representations from data, often using neural networks with many layers. RL, on the other hand, is focused on making decisions to achieve a goal, often in an interactive environment. However, the two can be combined in what's known as deep RL (DRL). In DRL, deep learning techniques are used to help the agent learn from its environment, enabling it to make better decisions. This combination has been particularly successful in complex tasks like playing Go and Poker, autonomous driving, and robotic control.

GENERATIVE AI

Generative AI represents a category of supervised, unsupervised, and deep learning models designed to generate new data that resembles an existing data set (Stokel-Walker & Van Noorden, 2023; Gozalo-Brizuela & Garrido-Merchan, 2023). While discriminative models aim to classify or differentiate between different kinds of data, generative models mainly **create**. They produce outcomes that extend beyond analyzing or categorizing data, venturing into the area of creation and simulation.

The term "generative AI" refers to ML models designed to generate new data that resembles a given dataset that can contain numerical textual or audiovisual data. These models learn the underlying patterns and distributions from the data during training and can produce novel outputs such as text, images, or music during inference. Popular examples include **GANs**, **VAEs**, and some types of **RNNs**.

If we ask ChatGPT to describe GANs using figurative language, we can get a description along these lines:

GANs, introduced by Ian Goodfellow and his colleagues in 2014 (Goodfellow et al., 2014), act like two opposing forces, a forger and a detective, caught in a continuous loop of creation and critique. The forger, known as the generator, attempts to produce realistic data, while the detective, the discriminator, strives to distinguish between the forger's creations and real data. This competition fuels a creative evolutionary process that continually refines the generator's ability to create convincing outputs.

So, in other words, GANs consist of two neural networks: the generator and the discriminator (Durall et al., 2021; Lee et al., 2022). The generator tries to produce data, while the discriminator tries to distinguish between real data and the data produced by the generator. As the two networks "compete" against each other, they both improve, leading the generator to produce increasingly convincing outputs.

On the other hand, VAEs, discovered by Diederik P. Kingma and Max Welling (Kingma & Welling, 2019), resemble a sculptor, carefully crafting and refining their creations. VAEs learn the underlying structure of the data and encode it into a compact form. They then generate new data by exploring variations within this learned structure, creating an output that is both novel and firmly grounded in the original data distribution. They work by compressing data into a lower-dimensional space and then reconstructing it. The "variational" aspect allows the model to generate new, varied outputs by exploring different points in this compressed space.

As we will see in this chapter, GANs and VAEs have carved niches for themselves, revolutionizing how we perceive and generate data. While GANs harness the adversarial battle between the generator and the discriminator to craft startlingly realistic data imitations, VAEs employ a probabilistic lens, focusing on latent space representations to capture and replicate data nuances. Their generative power offers a rich playground for applications ranging from art creation to simulating environments. Another important concept we need to introduce to get a better idea of the AI revolution is precisely **RL** that focuses on iterative decision-making through interactions with environments (Li, 2017). Although generative AI models like GANs and VAEs do not rely on RL, their training process has many elements of it that we will elaborate on later in this chapter. RL is gaining importance in generative AI and LLMs for a variety of reasons. It allows for more precise fine-tuning of models after their initial training, making them more effective at specific tasks. The algorithms can balance between exploring new strategies and exploiting known ones, which is crucial for generating creative yet coherent content. RL can also incorporate real-time feedback to adapt and improve, making the models more responsive to user needs. It's well suited for tasks in dynamic environments, such as conversational agents, and can be used to train models to adhere to ethical guidelines or safety constraints. As AI systems become increasingly autonomous, this is becoming more important. Additionally, as models grow in complexity, RL provides a scalable framework for training them. The combination of RL with LLMs also opens up new avenues for practical applications, from content generation to decision support systems.

In contrast, **transformers** are a type of neural network architecture introduced in 2017, primarily designed for handling sequence data. They are especially effective in NLP tasks and have become the foundation for many state-of-the-art ML models (Manning et al., 2014). Transformers use the self-attention mechanisms (that we will explain later) to weigh the importance of different parts of the input data, allowing them to capture long-range dependencies and contextual information. They are used in both generative and discriminative models. The transformer architecture has marked a shift in sequence-to-sequence learning. It revolutionized the booming field of NLP and birthed new models, including Claude or ChatGPT that facilitate our current discourse. Self-attention mechanisms of transformers adeptly discern relationships within sequences, capturing both local and distant contextual cues. This capability has positioned transformers as the foundation of applications in NLP. To offer a glimpse into the flexibility and capability of the transformers, we propose to dive right in and explore an alternative presentation style for this chapter, leveraging the capabilities of GPT – an LLM released by OpenAI (Fröhling & Zubiaga, 2021; King, 2023).

UNDERSTANDING GANS AND VAES FROM THE INSIDE

GENERATIVE ADVERSARIAL NETWORKS

GANs are a class of artificial intelligence algorithms used in unsupervised ML, implemented by a system of two neural networks contesting with each other in a game (in the sense of game theory, not video games). As we mentioned before, GANs consist of two parts: **the generator** and **the discriminator** (Liu & Hsieh, 2019).

The unique feature of this setup is the dynamic feedback loop at its heart. As the generator gets better at creating synthetic data, the discriminator must improve its ability to spot these fakes. In turn, as the discriminator gets better, the generator must improve its creations. Over time, this competition pushes the generator to create outputs that are almost indistinguishable from the real data.

This competitive dynamic makes GANs a powerful tool for generating new data. They have found extensive applications in areas like image synthesis, image super-resolution, and style transfer. The concept of GANs has also expanded, with variants such as deep convolutional GAN (DCGAN), Wasserstein GAN (WGAN), and cycle-consistent adversarial network (CycleGAN), each adding unique twists to the original model (Gulrajani et al., 2017; Wang et al., 2024). The potential of GANs extends beyond data generation. The underlying philosophy of GANs, the concept of adversarial learning, might provide us with an insightful analogy to human creativity and learning. Are we, too, shaped by adversarial forces? Do we also improve through competition and feedback? Could we learn something about our own creative processes from these artificial networks? We don't quite have the answers to these questions yet.

TYPES OF GANS

There are many versions of GANs, each with its unique characteristics, designed to solve specific problems or enhance certain aspects of the original GAN. Here are a few notable examples:

- **DCGAN:** This is one of the most popular types of GANs. Introduced by Alec Radford, Luke Metz, and Soumith Chintala in 2015, DCGAN employs convolutional neural networks (CNNs) in both the generator and the discriminator. It was one of the first models that stabilized the training of GANs and has since been widely used as a benchmark (Radford et al., 2015).

- **WGAN:** WGAN is a variant of GAN proposed by Martin Arjovsky, Soumith Chintala, and Léon Bottou in 2017 (Arjovsky et al., 06–11 August 2017). The key innovation in WGAN is the use of the Wasserstein distance, or Earth Mover's Distance, to measure the distance between the distribution of the generated images and the real images. This improves the stability of the GAN training process.

- **Conditional GAN (CGAN):** In this variant, both the generator and discriminator are conditioned on some extra information, such as a class label. This information can guide the data generation process towards a desired direction. CGANs, introduced by Mehdi Mirza and Simon Osindero in 2014 (Mirza & Osindero, 2014), can generate data with specified characteristics, making them particularly useful for tasks such as multimodal image synthesis.

- **CycleGAN:** Proposed by Jun-Yan Zhu and colleagues (Isola et al., 2016; Zhu et al., 2017), CycleGAN is a method for performing image-to-image translations without paired data. For example, it can transform a horse into a zebra or summer scenes into winter ones. It's named "CycleGAN" because it introduces a cycle-consistency loss to enforce forward-backward consistency, which improves the quality of the translated images.

- **StackGAN:** Introduced by Han Zhang and colleagues in 2017 (Zhang et al., 2017), StackGAN generates high-quality images from text descriptions. It does so in a two-stage process: the first stage generates a low-resolution image from the text description, and the second stage refines the low-resolution image into a high-resolution one (Figure 2.1).

These are just a few examples, and the world of GANs is constantly expanding with new types being developed to tackle specific challenges or to improve performance in certain applications. Each of these types offers unique insights into how the generative process can be controlled, guided, and refined, and collectively, they represent a significant advancement in the field of generative AI (Figure 2.2).

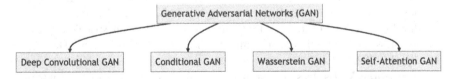

FIGURE 2.1   Examples of some of the GANs, diagram generated using ShowMe Plugin for ChatGPT.

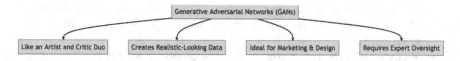

FIGURE 2.2   Examples of the features of GANs, diagram generated using ShowMe Plugin within ChatGPT.

VARIATIONAL AUTOENCODERS

VAEs constitute a class of generative models that have garnered attention for their unique approach to encoding and generating data. Rooted in the probabilistic graphical modeling paradigm, VAEs find their strength in representing data in a latent space, which allows for both the understanding of data structure and the generation of new samples (Doersch, 2016).

At a high level, VAEs can be visualized as comprising two principal parts: an encoder and a decoder. The encoder maps input data to a latent space, while the decoder maps points in the latent space back to the data space. However, instead of deterministically mapping data to a single point in the latent space as traditional autoencoders might, VAEs introduce a probabilistic aspect. The encoder outputs parameters of a probability distribution, typically Gaussian, from which we can sample multiple potential latent representations for a given data point.

The innovation of VAEs comes in the form of a specific type of loss function. This loss is a combination of two components: a reconstruction loss, which ensures the decoded samples resemble the original data, and a regularization term, which forces the latent space to adhere to a predefined distribution (usually a standard normal distribution). This design enables the generation of new, coherent samples by merely sampling from the latent space and passing those samples through the decoder.

Over time, several variants and improvements upon the basic VAE framework have been proposed. A few of these include the following:

- **Conditional VAEs (CVAEs):** These are an extension to VAEs that allow for the generation of data samples conditional on certain variables. By introducing auxiliary information during training, CVAEs can generate data samples based on specific conditions or attributes (Sarkar & Cooper, 2021).
- **Beta-VAEs:** This variant introduces a hyperparameter, Beta, to the VAE loss function. By controlling the weight of the regularization term in the loss, Beta-VAEs offer a trade-off between the latent space's expressiveness and its alignment with the predefined distribution (Hall & Sparks, 2023).
- **Sequential VAEs:** Designed to handle sequences, this type of VAE introduces recurrent structures, like LSTM or Gated Recurrent Unit (GRU) cells, into the encoder and decoder parts, allowing them to process time series or sequential data more effectively (Geenjaar et al., 2022).

- **Disentangled VAEs**: The goal here is to represent the latent variables in a way where individual dimensions are more interpretable and less correlated. This leads to better control and understanding of the generative factors in the data (Poels & Menkovski, 2022).

Each of these variants caters to specific requirements or challenges associated with different data types or desired model properties. By understanding the underlying principles and the extensions brought about by these variants, one can tailor VAEs to a multitude of tasks, ranging from image synthesis to anomaly detection (Figure 2.3).

COMPARISON BETWEEN VAEs AND GANs

Now, how are VAEs technically different from GANs? VAEs and GANs are both powerful generative models that can generate new data that resembles the input data they have been trained on. However, the way they learn to generate data and the properties of the data they generate are quite different. Let's compare them based on their structure, learning approach, and the characteristics of the generated data.

Structure:

- GANs consist of two neural networks – a generator and a discriminator – that play a competitive game. The generator creates fake data to pass off as real, and the discriminator tries to distinguish between the real and the fake.
- VAEs, on the other hand, are structured as an encoder-decoder pair. The encoder network reduces the input data into a lower-dimensional representation, and the decoder network then reconstructs the original data from this compressed representation.

Learning Approach

- GANs learn via a zero-sum game where the generator tries to fool the discriminator, and the discriminator tries not to be fooled. This adversarial process can lead to highly realistic data, but the training process can be unstable and might not converge.

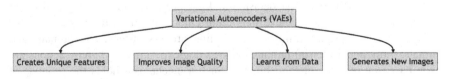

FIGURE 2.3 Examples of some of the VAEs, diagram generated using ShowMe Plugin for ChatGPT.

- VAEs learn by maximizing the likelihood of the generated data being similar to the input data while also maintaining a smoothly varying, low-dimensional latent space. This makes the training process of VAEs more stable than GANs, but it can sometimes lead to generated data that are blurrier than those produced by GANs.

Characteristics of Generated Data

- GANs often produce data that closely resemble the training data, which makes them particularly useful for tasks that require the generation of high-quality, realistic samples.
- VAEs, due to their probabilistic nature and the enforced continuity in the latent space, are better at ensuring the diversity of generated data and are more suitable for tasks that require smoothly varying output or interpolation between data points.

While both GANs and VAEs are capable of generating new data, they each have strengths and weaknesses that make them suited for different tasks. Understanding these differences can help guide the choice of model for a particular application and can also shed light on the different aspects of the creative process that these models embody.

THE ROLE OF RL AND ACTIVE LEARNING IN COLLABORATIVE AI

Now let us dig into RL. The popularity of RL can be attributed to the unique value proposition it brings to the table, which fundamentally differs from traditional supervised and unsupervised learning paradigms (Sutton and Barto, 2018).

RL represents a distinct paradigm within the ML landscape, characterized by an agent (autonomous entity that acts upon an environment to achieve its goals (Steels and Brooks, 2018)) that engages with its environment through actions, gauges the consequences via feedback, and subsequently refines its strategy. At its core, RL is concerned with making decisions. In this type of ML an agent learns to make a sequence of decisions by interacting with an environment to achieve a goal or maximize some notion of cumulative reward. RL has roots in operations research, behavioral psychology, and control theory. Most importantly though, in contrast to older ML approaches, like supervised learning, where explicit correct answers guide the learning process, RL discovers answers predominantly on its own (Sewak, n.d.; Arel, 2012).

At its core, RL focuses on decision-making processes, optimizing actions in dynamic environments to achieve maximum cumulative rewards over time. This decision-centric approach mimics the learning process observed in humans and animals, think about the (in) famous Palvov's dog experiment, whereby actions are taken based on rewards or punishments. Such a mechanism holds the promise of creating AI agents that can learn from interactions, adapt over time, and eventually make decisions.

The agent operates within a designated environment and acts based on the current state it perceives. The aftermath of this action is relayed back to the agent in the form of rewards, which serve as a measure of the action's efficacy. The quintessence of RL revolves around the agent's pursuit to master a policy, essentially a function that suggests optimal actions for every conceivable state, to optimize the expected cumulative rewards across its operational lifetime.

The ascendancy of RL in recent academic and practical circles is attributable to multiple intertwined factors. Foremost, RL has showcased an unparalleled capacity to model decision-making processes that have ramifications extending deep into the future. Such capabilities have serious implications, spanning a spectrum of applications, from digital entertainment, such as video games, to the tactile world of robotics and the abstract domain of finance (Mnih et al., 2013). The wider recognition of RL's potential received a significant impetus with landmark achievements such as DeepMind's AlphaGo, which in 2016 demonstrated a capacity in the game of Go surpassing human champions (Holcomb et al., 2018; Bory, 2019).

Recent innovations that amalgamate deep learning with RL have further bolstered the stature of RL. The adeptness of deep neural networks in processing voluminous data and approximating complex functions facilitates the application of RL to multifaceted, real-world tasks that traditionally resisted computational modeling. Finally, the expanding horizon of RL applications, encapsulating fields as diverse as digital advertising optimization, supply chain enhancements, and scientific research assistance, underscores its transformative potential in contemporary computational research (Polydoros & Nalpantidis, 2017; Coronato et al., 2020; Font & Mahlmann, 2021).

INTERPLAY OF RL WITH GANS AND VAES

GANs and VAEs are predominantly recognized for their generative capabilities, with an intrinsic aim to learn and replicate specific data distributions. Conversely, RL traditionally anchors itself in the domain of optimizing sequential decision-making. However, a deeper introspection into the mechanics and methodologies underlying these paradigms reveals a web of connections, symbiotic relationships, and potential collaborative avenues (Atienza, 2020).

The generative essence of GANs and VAEs has found synergies with RL in scenarios where data simulation is important. For instance, the deployment of GANs for simulating environments or creating synthetic data bridges a significant gap in RL, especially in contexts where real-world data acquisition poses logistical or financial challenges. This simulated data, bearing high fidelity to actual observations, can then serve as a fertile training ground for RL agents (Babcock & Bali, 2021).

Furthermore, the paradigm of model-based RL, wherein the agent endeavors to discern or approximate its environment's model, invites integration with generative models like GANs and VAEs. These generative structures can be tasked with simulating prospective environmental states, equipping the RL agent with the foresight to anticipate the ramifications of its chosen actions, thereby enabling informed decision-making. Concomitantly, RL itself can be reciprocally employed to enhance the training processes of GANs and VAEs. By deploying RL as an optimization tool, the generative processes, particularly within GANs, can be fine-tuned to produce outputs that not only are of superior quality but also exude stability across iterations.

TRANSFORMERS AND RNNS: A COMPARATIVE LENS ON SEQUENCE MODELING

Transformers and RNNs are two architectures that often come into focus in the world of NLP and sequence modeling. They also serve as the base of LLMs like the ones we used to help us write this book. While they serve similar purposes in handling sequence data, their approaches, complexities, and efficiencies differ significantly. Understanding these differences is crucial for grasping how advancements in ML and deep learning have shaped the capabilities of modern natural language models, including LLMs like GPT-4 (Sanderson, 2023; Katz et al., 2023).

RNNs have been a cornerstone in sequence modeling for years. The fundamental idea behind RNNs is the incorporation of "memory" into neural networks (Carpenter, 1989; Yao et al., 2014; Hochreiter & Schmidhuber, 1997). Unlike standard feedforward neural networks, which treat each input independently, RNNs maintain a hidden state that captures information about previous steps in a sequence. This ability allows RNNs to excel in tasks that require an understanding of context or the sequential nature of data, making them useful for time-series analysis, speech recognition, and early NLP tasks. While revolutionary in their time, RNNs have limitations, particularly concerning computational efficiency and the ability to capture long-term dependencies in a sequence. Training an RNN involves a lot of repetitive computation, which can be computationally expensive and time-consuming. The architecture also suffers from problems like vanishing and exploding gradients, which make it difficult for the network to learn from earlier parts of a sequence – the so-called "long-term dependencies" (Salehinejad et al., 2017).

Transformers were introduced to overcome some of these limitations, initially as a mechanism for machine translation. They are a type of neural network architecture introduced in 2017, primarily designed for handling sequence data. They are especially effective in NLP tasks and have become the foundation for many state-of-the-art ML models. Transformers use self-attention mechanisms to weigh the importance of different parts of the input data, allowing them to capture long-range dependencies and contextual information. They are used in both generative and discriminative models. The architecture got rid of recurrence altogether and relied heavily on attention mechanisms to draw global dependencies between the input and the output. This allowed transformers to process all words or tokens in the sequence simultaneously, leading to significantly reduced training times and improved performance on capturing long-term relationships in the data.

Both RNNs and transformers aim to understand the structure in sequences of data, but they go about this in different ways. While RNNs loop through each element in a sequence one at a time to capture contextual information, transformers look at an entire sequence at once, using attention mechanisms to weigh the importance of different parts of the input data. This makes transformers particularly well suited for tasks that require understanding the entire context, leading to their adoption in LLMs and other advanced NLP tasks (Du et al., 2021).

As we explore the role of advanced ML models such as LLMs in enhancing human-machine collaboration (Dobrin, 2023; Hendry et al., 2023), understanding the transition from RNNs to transformers is vital. Transformers have essentially become the backbone of modern NLP and a key component in LLMs. They encapsulate the advances made in deep learning and sequence modeling, overcoming many of the challenges that plagued earlier architectures like RNNs.

RNNs and transformers are both milestones in the evolving journey of sequence modeling. While RNNs laid the foundational work for understanding sequences, transformers have built upon and transcended those foundations, becoming an integral part of the state-of-the-art models that are pushing the boundaries of what's possible in natural language understanding and generation.

TRANSFORMERS AND THEIR SYMBIOSIS WITH RL

Understanding the relationship between RL and deep learning is crucial, especially when we consider the development of complex systems like LLMs. While LLMs primarily leverage deep learning techniques, especially transformers, for natural language understanding and generation, there's growing interest in incorporating RL into these models (Ozdemir, 2023). For example, RL techniques could train an LLM to better interact with humans in a conversation by optimizing some reward function that measures the quality of the dialogue.

As mentioned already, the transformer architecture (Hogan, 2023), at its core, is distinguished by its self-attention mechanism. This mechanism allows the model to weigh the relevance of different parts of an input sequence in relation to a particular point in the sequence. By doing so, transformers can discern relationships and dependencies between disparate elements, capturing both local and long-range contextual information. This capacity to dynamically adjust the focus on different parts of the input, depending on the context, lends the architecture its powerful capability to handle sequences.

RL, on the other hand, is fundamentally about learning by interaction. Agents take actions in environments and receive feedback in the form of rewards or penalties. The objective is to discover policies that maximize some notion of cumulative reward over time. This iterative process of action and feedback allows agents to refine their strategies, adapting to complex, often unknown, environments.

The confluence of transformers and RL emerges from their complementary strengths. Transformers, with their capability to model complex relationships in data, can serve as powerful function approximators within the RL framework (El Sallab et al., 2017). When an RL agent needs to process sequences of observations, as is the case in many real-world scenarios, the transformer can encode these sequences into meaningful representations that capture the salient features and temporal dependencies essential for decision-making.

For instance, in tasks that involve natural language instructions or multimodal inputs, where the agent must integrate visual information with textual descriptions, the transformer's capacity in processing sequences can be harnessed to generate richer state representations. These representations, when fed into an RL algorithm, can facilitate more informed policy decisions, leading to enhanced performance.

Furthermore, the very architecture of transformers (Devlin et al., 2018) lends itself to exploration in RL. The self-attention mechanism can be interpreted in the context of RL as a way of dynamically allocating attention based on the importance of different observations, similar to how an agent might weigh the significance of different states or experiences.

In many ways, the integration of transformers into RL scenarios exemplifies the broader trend in ML research: the fusion of powerful representational models with algorithms designed for decision-making under uncertainty. Such interdisciplinary engagements offer a glimpse into the future of artificial intelligence, where models adept at understanding patterns in data collaborate with decision-making frameworks, leading to agents that can both comprehend and act adeptly in complex environments.

SUMMARY: POINTS OF INTERSECTION

Natural Language Understanding for RL: Transformers can be employed to provide natural language understanding capabilities to RL agents. For example, an RL agent navigating a text-based environment can use a transformer to understand instructions or interpret text descriptions, thereby making more informed decisions.

Sequence Decision-making: Some RL problems are essentially about making a sequence of decisions over time. Transformers, which excel at handling sequences, can be used to model the state space or the policy in RL, especially when the state or action space has a sequential or structured nature.

Meta-Learning and Transfer Learning: Transformers pretrained on massive datasets can be fine-tuned using RL for specific tasks, combining the general language understanding of transformers with the task-specific optimization of RL.

Explainability and Interpretability: The attention mechanisms in transformers can be leveraged to understand the decision-making process of an RL agent, potentially offering insights into why an agent took a particular course of action.

Optimization of Reward Functions: In some settings, transformers can be used to learn the reward function for an RL agent based on natural language descriptions or demonstrations, aligning the agent's behavior with human-like objectives.

LLMs ON THE RISE

As we dwell into generative AI, and more specifically, algorithms that allow us to process and generate text, the time is now to turn our attention to LLMs like

GPT-4 (Huck, 2023). These models embody the true epitome of generative AI within NLP, showcasing synthesis of deep learning architectures and generative algorithms.

The mechanisms that underlie LLMs are rooted in both deep learning and generative AI. LLMs often employ transformer architectures, a paradigm-shifting deep learning model initially designed for translation tasks. Over time, the versatility and effectiveness of transformers have been harnessed to create increasingly sophisticated language models that can generate humanlike text, understand context, and even reason to a certain extent. This places them squarely within the domain of generative AI, as their primary function is to produce new, coherent, and contextually relevant text based on the data they have been trained on.

What sets LLMs apart is their ability to traverse the pathways of human language, understanding not just syntax and semantics but also nuances, idioms, and cultural contexts. The LLMs go beyond data generation and enter the stage of "understanding," at least in a computational sense. They are at the forefront of technology where machine capabilities start to mimic human cognitive processes related to language, such as summarization, translation, question-answering, and even elements of common sense reasoning.

While GANs may be painting digital images and VAEs may be mapping data distributions, LLMs contribute to the arts, sciences, and industries by "speaking" and "writing," by generating text that can inform, entertain, persuade, or even emulate humanlike conversation. They serve as an interface where humans can interact most naturally with the machine which is through language. The essence of LLMs, therefore, is not just algorithmic but also linguistic, creating a seamless continuum between human communicative capabilities and machine computational power.

LLMs represent an arena where deep learning's computational capacity and generative AI's creative potential converge most naturally with human intellectual and communicative faculties. This synthesis underscores the power of individual technologies and highlights the synergistic possibilities that arise when these technologies are aligned in harmony with human skills and needs.

Some of the most popular LLMs include the following:

- GPT-3 (Generative Pre-trained Transformer 3) is a commercial LLM developed by OpenAI. It was released in 2020 and has approximately 175 billion parameters (Floridi & Chiriatti, 2020).
- GPT-4 is the latest version of GPT-3, released in 2022. It has over 175 billion parameters, but it is also more efficient and can generate text faster than GPT-3 (Koubaa, 2023).
- PaLM (Pathway Language Model) is a commercial LLM developed by Google AI. It was released in 2022 and has over 540 billion parameters. PaLM is capable of a wide range of tasks, including text generation, translation, summarization, and question answering (Peng et al., 2019).
- Codex is a commercial LLM developed by Google AI. It was released in 2022 and is capable of generating code in a variety of programming languages (Finnie-Ansley et al., 2023).

- Cohere is an open-source LLM developed by Cohere. It was released in 2022 and has over 100 billion parameters. It is capable of generating text, translating languages, and answering questions (Huang et al., 2023).
- LaMDA (Language Model for Dialogue Applications) is an open-source LLM developed by Google AI. It was released in 2021 and is designed for dialogue applications (Martin, 2023).
- BERT (Bidirectional Encoder Representations from Transformers) is an open-source LLM developed by Google AI. It was released in 2018 and is used for natural language understanding tasks such as text classification and question answering (Devlin et al., 2018).
- Transformer-XL (Transformer-Xtra Long): Developed by researchers at Google Brain, this model improves upon the original transformer architecture by introducing a recurrence mechanism to capture longer-term dependencies. It was released in 2019.
- XLNet: Created by researchers at Carnegie Mellon University and Google Brain, XLNet is an extension of Transformer-XL and was designed to outperform BERT in several benchmarks. It was released in 2019.
- T5 (Text-To-Text Transfer Transformer): Developed by Google AI, T5 interprets all NLP tasks as text-to-text tasks, simplifying the process of applying the model to diverse tasks. It was released in 2019.
- RoBERTa (Robustly Optimized BERT Pretraining Approach): Developed by Facebook AI, RoBERTa is a variant of BERT that is optimized for more robust performance. It was released in 2019.
- GPT-2: Developed by OpenAI, it is the predecessor to GPT-3. It was released in 2019 and has 1.5 billion parameters. It was notable for its text generation capabilities but was initially not fully released due to ethical concerns.
- ERNIE (Enhanced Representation through kNowledge IntEgration): Developed by Baidu, ERNIE is designed to better understand the semantics of words by considering their relationship with other words. It was released in 2019.
- ELECTRA (Efficiently Learning an Encoder that Classifies Token Replacements Accurately): Developed by researchers at Stanford and Google, ELECTRA is designed to be more sample efficient than models like BERT. It was released in 2020.
- DALL-E: Developed by OpenAI, DALL-E is a GPT-3 variant trained to generate images from textual descriptions. Though not strictly a text-based LLM, it's an interesting extension of the technology. It was released in 2021.
- CLIP (Contrastive Language-Image Pretraining): Another model by OpenAI, CLIP is designed to understand images in context with natural language descriptions. Like DALL-E, it's an extension of LLMs into the visual domain. It was released in 2021.
- JurisPRLM: Developed by Thomson Reuters, it is a specialized LLM for legal texts. It aims to assist in legal research and drafting. It was released in 2021.

- Claude: Developed by Anthropic, Claude is a next-generation AI assistant. It has been tested in partnership with companies like Notion, Quora, and DuckDuckGo. Unique features include its high reliability, steerability, and the ability to customize its personality and tone. It addresses some of the limitations of existing LLMs.

The history of LLMs can be traced back to the early days of ML. In the 1950s, researchers developed simple language models that could predict the next word in a sentence. However, these models were not very powerful and could not handle complex tasks.

In the 1990s, researchers developed more powerful language models based on neural networks. These models were able to achieve better performance on a variety of tasks, but they were still limited by the amount of data they could be trained on. In recent years, the development of LLMs has been accelerated by the availability of massive datasets of text and code. These datasets have allowed researchers to train models with billions of parameters, which has led to a significant improvement in performance. Also, the future of LLMs is very promising. These models have the potential to revolutionize a wide range of industries, including healthcare, education, and customer service. However, there are still challenges that need to be addressed before LLMs can be even more widely used across different domains. One challenge is that these models can be biased, reflecting the biases that are present in the data they are trained on. Another challenge is that these models can be used to generate harmful content, such as hate speech or misinformation. Despite these challenges, LLMs are a powerful new technology with the potential to change the world. As these models continue to develop, it is important to be aware of their limitations and to use them responsibly.

TRAINING THE TITANS: GANS, VAES, AND TRANSFORMERS

GANs, VAEs, and transformers offer unique strengths but all rely on iterative refinement for training. This chapter explores their training mechanisms, from GANs' adversarial balance (Gulrajani et al., 2017) to VAEs' dual-objective focus on reconstruction and regularization (Miladinović et al., 2022), and transformers' attention-based handling of sequential data.

During the training of GANs, this adversarial dance becomes progressively refined. The generator improves its ability to produce convincing data, while the discriminator hones its discernment. Through multiple interactions, a balance is struck where the generator crafts data almost indistinguishable from real samples, and the discriminator, in its heightened acuity, finds it increasingly challenging to differentiate (Gulrajani et al., 2017).

For VAEs, two principal components come into play: reconstruction loss and regularization loss. The former ensures the decoded output closely mirrors the original input, while the latter encourages the latent space to adhere to a predefined distribution, typically a Gaussian. This dual-objective training ensures that the model not only recreates data effectively but also captures the underlying data distributions, allowing for meaningful data generation and manipulation (Miladinović et al., 2022).

Born out of the necessity to process sequential data more effectively, transformers revolutionize sequence-to-sequence tasks without relying on recurrence. By assigning different attention scores, transformers can capture long-range dependencies in data, something traditional recurrent networks often struggled with. During training, the model adjusts these attention scores and other associated weights to optimize the prediction of the next element in a sequence, be it a word, a note, or any other data point. The beauty of transformers, especially in larger incarnations, is their capacity to generalize across tasks, as they learn universal patterns from vast amounts of data (Wang et al., 2023).

In each of these methodologies, the underlying principle remains consistent: iterative refinement. Through cycles of forward and backward propagation, and the subtle tweaking of thousands to millions of parameters, these architectures learn to model, generate, and process data in ways that are pushing the boundaries of what ML can achieve. Their training and refinement processes, while diverse in nature, underscore a relentless pursuit of accuracy, efficiency, and adaptability.

TRANSFORMERS: UNRAVELING THE SELF-ATTENTION MECHANISM

The impact of the transformer architecture on modern ML, particularly in NLP and sequence-to-sequence tasks (Tian et al., 2022), can be largely attributed to its unique training methodology. Its departure from recurrent and convolutional paradigms propelled a novel approach to sequence (Cui et al., 2019). Training a transformer entails adjusting numerous parameters to optimize the model's capacity to predict or generate subsequent elements in a sequence. Every layer in the transformer model carries its set of learnable parameters (Alekseev & Bobe, 2019). As input data traverses these layers, internal representations evolve, with each layer capturing progressively more abstract and complex relationships. The central point of training, then, is to fine-tune these parameters such that the final output aligns closely with the expected outcome. This is achieved by minimizing a loss function, which quantifies the disparity between the model's predictions and the actual data. Since the transformer lacks inherent recurrence, it cannot intrinsically perceive the order of elements in a sequence. Positional encodings resolve this by embedding information about element positions into their respective embeddings before they undergo self-attention (Wu et al., 2019). These encodings, usually sinusoidal functions, ensure that the model remains aware of element positions throughout its depth, facilitating more accurate contextual interpretations.

Given their vast number of parameters, especially in architectures like BERT or GPT variants, transformers are susceptible to overfitting (Ying, 2019). Regularization techniques, such as dropout, where random subsets of neurons are "dropped" or turned off during training, play a significant role in ensuring the model generalizes well to unseen data. Layer normalization, another crucial component, aids in stabilizing the activations across layers, fostering smoother and faster convergence during training (Ba et al., 2016).

Moreover, the training of transformer models often employs adaptive learning rate algorithms, such as the Adam optimizer (Zhang, 2018) (Tato & Nkambou, 2018). These algorithms adjust learning rates dynamically based on the evolution of gradients, ensuring that the model converges to a solution efficiently without oscillating or overshooting. As the transformer undergoes training, its ability to discern patterns, relationships, and nuances in the data augments. Each pass, each epoch, refines its internal weights, making it progressively more adept at its task. The culmination of this process is a model that can navigate the sequences, from languages to music, with a depth and finesse that has set new benchmarks in the world of ML.

TRAINING, FINE-TUNING, AND PROMPTING: THE TRIAD SHAPING LLMS

When engaging with LLMs, such as GPT and similar architectures, it is important to recognize the multistaged evolution of the model from its inception to its interaction. There are three cardinal stages that govern this trajectory: training, fine-tuning, and prompting. These stages interweave and overlap in their roles, each essential to the manifestation of the model's capabilities:

- **Knowledge Transfer (TRAINING):** The foundational knowledge acquired during training sets the stage for fine-tuning. Without solid foundational training, fine-tuning becomes an exercise in futility, akin to attempting specialized studies without primary education (Cassano et al., 2023).
- **Iterative Refinement (FINE-TUNING):** The prompting stage can sometimes reveal shortcomings or blind spots in the model's responses. This feedback can loop back into the fine-tuning stage, leading to iterative refinement of the model (Madaan et al., 2023).
- **Prompting as Adaptive Learning (PROMPTING):** Every interaction, though primarily extraction, is also a form of real-time adaptive learning. The model doesn't retain new knowledge but refines its response mechanism based on the prompts it encounters (Wan et al., 2023) (Figure 2.4).

In the larger scheme, training equips the model with a wide-angle lens to view the world, fine-tuning zooms into specific areas of interest, and prompting ensures the focus remains sharp and the picture clear. Together, they harmonize to ensure LLMs are both knowledgeable and tractable, a combination that's crucial for their wide-ranging applications in academia, business, and beyond (Kampik et al., 2023).

Training, the foundational phase, involves immersing the LLM in extensive corpuses of text (Li et al., 2023; Peng et al., 2023), enabling it to capture linguistic patterns, structures, and intricacies of human language. By adjusting its internal weights and biases in response to vast troves of data, an LLM learns to generate coherent, contextually appropriate responses. At this stage, the architecture of the model assimilates the generic properties of language, making it a tool for a large

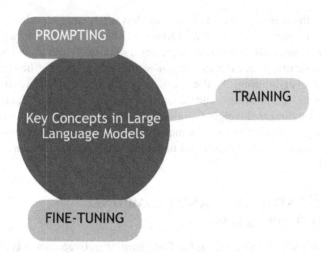

FIGURE 2.4 The mind map provides a visual overview of three key concepts in LLMs: Training, fine-tuning, and prompting created with diagrams plug-in within ChatGPT.

number of linguistic tasks. At the heart of any LLM's capabilities lies its foundational training. This is the stage where the neural networks are exposed to a deluge of textual data. This vast information ocean has waves of grammar, currents of logic, and tides of reasoning that the model learns to navigate. Training is the molding of raw potential. A well-trained model has an extensive general knowledge base and can draw upon this wealth of information to construct or understand language sequences. It's analogous to a person immersing themselves in years of general education. However, it's essential to acknowledge that training is also where biases embedded in data are inadvertently absorbed by the model. The model's neutral stance means it is influenced by the dominant narratives present in its training data. The initial training phase is where an LLM ingests vast swathes of data. This foundational training is similar to a child undergoing their primary education, acquiring a general understanding of a spectrum of topics. The model learns language structure, facts, opinions, reasoning abilities, and biases present in the data. By optimizing weights in the neural network, the model learns to predict the next word in a sequence, refining its internal representation of the language (Aljarah et al., 2018).

The sheer breadth and diversity of data introduced during training can sometimes lead to a lack of precision in niche domains or specific tasks. Once the base model is trained, it is often not immediately deployed for tasks. It undergoes a secondary phase called fine-tuning. During this phase, the model is exposed to more specific datasets, usually tailored to particular tasks or domains. It's similar to a student pursuing higher studies in a chosen field after a broad-based schooling. Fine-tuning sharpens the model's capabilities, making it adept at specialized tasks. It also offers an opportunity to reduce or realign any biases and align the model's behavior more closely with desired outcomes. This subsequent phase narrows down the model's focus to task-specific datasets, allowing it to acquire expertise in a targeted domain without being swayed by the vast general knowledge acquired during training. By refining the model's weights further on this curated data, fine-tuning ensures that the LLM can display both depth and specificity in its responses, becoming an expert in narrower fields. If training is about forging the metal, fine-tuning is the art of turning that forged metal into a precise instrument. The fine-tuning datasets, typically smaller and more task-specific, allow the model to home in on nuances and subtleties of particular domains or applications. Think of it as a researcher diving into specialized courses after a broad undergraduate study. It's here that the model's behavior is aligned more closely with human-defined ideal outputs. This phase can also act as a corrective mechanism where efforts can be made to counteract or minimize the biases absorbed during the training (Yu et al., 2023).

Prompting, on the other hand, is a nuanced technique to harness the knowledge and adaptability of the LLM without further adjusting its internal architecture. By crafting precise, evocative prompts, users can guide the LLM to retrieve or generate specific information or content, effectively leveraging the model's expansive knowledge base. The art of prompting, thus, capitalizes on the model's intrinsic ability to generalize from its training and fine-tuning phases, offering a dynamic way to access its vast reservoirs of information.

In essence, while training lays the foundational linguistic knowledge, fine-tuning sharpens the model's ability in specific domains. Prompting, as a strategy, then navigates this vast and nuanced knowledge pool, ensuring that the LLM's responses are both relevant and insightful. Post-training and fine-tuning, the model becomes a repository of patterns and vast knowledge. However, extracting the desired output from this repository is an art. Prompting becomes the bridge between a user's query and the model's potential answers. Through crafted prompts, users can guide the LLMs' attention towards specific areas of its knowledge base. Techniques like zero/few-shot learning (Chen et al., 2023) allow the model to generalize from a few examples, while chain-of-thought (CoT) or ReAct techniques ensure coherent, context-aware responses.

With a trained and often fine-tuned model in place, the next challenge is interaction. How does one extract relevant information or ensure the model behaves desirably? This is where prompting comes into play. Prompting is not just about posing questions; it's an art of guiding the model's vast knowledge and capabilities. Through techniques like zero/few-shot learning, CoT (Wei et al., 2022), or ReAct (Dagan et al., 2023), users can shape the model's outputs, probing deeper into topics or ensuring a coherent line of reasoning. It's analogous to having a conversation with an expert where the nature and phrasing of questions can significantly influence the depth and direction of the discussion.

In essence, training, fine-tuning, and prompting form a continuum in the life of an LLM. While training and fine-tuning shape the model's knowledge and capabilities, prompting refines the art of accessing and channeling this power during interactions. Together, they represent the choreography of steps that lead to the effective deployment and utilization of LLMs in various applications.

CHANNELING LLMS: THE NUANCES OF PROMPTING TECHNIQUES

In the evolving discipline of ML, particularly with the advent of sophisticated models like LLMs, the task of harnessing their vast capabilities has assumed paramount importance. Unlike traditional models, where the main challenge was attaining accuracy or efficiency, with LLMs, the central concern often revolves around directing their knowledge and inference power appropriately. Among the tools employed to this end are various prompting techniques, each bringing a unique flavor to the interaction with LLMs.

Zero/Few-Shot Learning

The terms "zero-shot" and "few-shot" are indicative of the amount of task-specific training data that the model is exposed to. In the area of LLMs, these terms often pertain to how the model is prompted:

- **Zero-Shot Learning:** In this paradigm, the LLM is provided with a task it has never explicitly been trained on. It relies on its vast general knowledge to decipher the prompt and generate a relevant output. This is similar to asking a well-read individual a question on a topic they haven't studied formally; their broad knowledge might still allow them to provide a meaningful answer (Chai, 2022).
- **Few-Shot Learning:** Here, the LLM is given a few examples (or "shots") of a task before presenting it with a new instance. The idea is to nudge the model into recognizing the pattern or context from these examples and then generalize to the new prompt (Akiti, 2021). Few-shot learning is increasingly used in medical imaging to identify rare conditions or anomalies. Traditional ML models require large datasets to train on, but in the case of rare diseases, only a few examples may be available. Few-shot learning algorithms can be trained to make accurate diagnoses based on these limited samples, thereby aiding healthcare professionals in quicker and more accurate decision-making.

CoT/Self-Consistency:

CoT, or chain-of-thought, prompting is a technique designed to sustain context or a line of reasoning across multiple interactions with the model (Iliadis et al., 2023):

- The primary idea behind CoT is to frame prompts in a manner that makes the model consider previous interactions or information, effectively creating a chain of interlinked responses.
- This approach encourages the model to exhibit self-consistency, wherein its outputs adhere to a logical and coherent line of reasoning, even across extended interactions.

ReAct: Recursive Prompt Amplification:

ReAct, short for Recursive Prompt Amplification (Bhargava & Sharma, 2021), is a novel approach that capitalizes on the recursive nature of interactions:

- In this paradigm, an initial prompt is given to the LLM. Once a response is generated, this response (either in entirety or selectively) is then repurposed as a part of a new prompt. The process can be iterated multiple times.
- The core advantage of ReAct is its potential to dive deeper into a topic or line of thought, allowing users to extract more nuanced or detailed information from the LLM.

In harnessing the expertise of LLMs, these prompting techniques serve as conduits, each shaping the flow of information in distinctive ways. While the underlying model remains consistent in its vast knowledge and capabilities, the choice of prompting strategy can significantly influence the depth, direction, and detail of its outputs.

USE CASES

AUTOMATED CONTENT SUMMARIZATION FOR NEWS WEBSITES USING TRANSFORMER

In the fast-paced world of digital journalism, readers often grapple with information overload. While they wish to stay informed, sifting through lengthy articles can be time-consuming. News websites and digital media platforms, recognizing this challenge, aim to provide concise summaries for their articles to offer readers a quick overview. To automate this summarization process, many have turned to the transformer architecture, given its proficiency in understanding context and generating coherent text.

Imagine a news website that publishes dozens of articles daily, ranging from international politics to local sports events. For each of these articles, a transformer-based model can generate a brief, coherent summary, capturing the essence of the content. When a journalist submits their article to the content management system, the model immediately crafts a summary which can either be used as-is or serve as a starting point for editors.

The self-attention mechanism of the transformer is crucial here. In a detailed article discussing, for instance, a recent political summit, there might be multiple themes: the main agenda of the summit, key participants, major decisions made, and perhaps some controversies or sideline events. A naive summarization might miss out on these nuances. However, the transformer, by weighing the importance of different segments of the article through its self-attention layers, can generate a summary that encapsulates the main points while preserving the article's overarching context.

For readers, this means that upon visiting the news website, they can quickly glance through the summaries, gaining an understanding of the day's major events. If a particular story piques their interest, they can look into the full article. For the news platform, this could translate to higher reader engagement, reduced bounce rates, and an enhanced user experience. By leveraging the transformer architecture, digital media platforms can cater to the modern reader's needs, ensuring that crucial information is accessible at a glance while maintaining the depth and richness of journalistic content.

PERSONALIZED CONTENT GENERATION FOR E-COMMERCE PLATFORMS USING VAEs

In the vast expanse of e-commerce, personalization has emerged as a potent tool to enhance user experience and boost sales. Customers today don't just seek products; they desire experiences tailored to their preferences and needs. Here, VAEs come into play, particularly in personalized content generation. Picture an e-commerce platform that hosts an extensive range of products, from fashion apparel to tech gadgets. When a user logs in, the platform's goal is to present items and promotions most relevant to that individual user, based on their past behavior and latent needs.

Using VAEs, the platform can craft personalized product images. The idea is not just to show a generic image of a product but to display it in a context resonating with the user. For instance, for a user who has shown interest in hiking, a pair of shoes can be showcased against a backdrop of mountains or trails, subtly emphasizing their suitability for hiking. VAEs achieve this by first learning a latent representation of the vast array of product images and user interactions available on the platform. As a generative model, a VAE can then produce new images by sampling from this latent space. By aligning this with the user's profile and interaction history, the model generates images tailored to each user's inclinations.

Personalization extends beyond static images. VAEs can assist in generating personalized video content for products, showing them in action in a manner most likely to resonate with the viewer. A tech gadget, for example, can be showcased in different settings – for a traveler, it might be shown being used in various global destinations, while for a student, in study or campus settings.

Such deep personalization fosters a connection between the user and the product, making them more likely to engage, explore further, and potentially make a purchase. This application transforms the typically generic e-commerce browsing experience into something uniquely catered to each visitor, showcasing the immense potential of VAEs in shaping the future of online retail.

DIVERSE PRACTICAL APPLICATIONS OF GANs

GANs, comprising two neural networks engaging in an adversarial process, have laid the groundwork for innovative applications across diverse industries. GANs are currently starting to play a crucial role in sectors like entertainment and design, aiding in the creation of digital art, music, and in-game assets. Their capacities extend beyond content creation, serving as invaluable tools for data augmentation in scenarios where data might be scarce.

Retail and fashion domains are witnessing the utility of GANs in visualizing and prototyping new designs, and automating the generation of varied product images. Virtual prototyping and automated image generation have become integral in expediting the design phase and creating a more engaging online shopping experience. Personalized marketing campaigns, driven by GAN analysis of customer preferences, are enhancing engagement and boosting sales.

In healthcare, GANs can help synthesize realistic images when data is scarce. These networks contribute to disease prediction, offering insights into preemptive care and treatment planning. Real estate and architecture utilize GANs for virtual staging of properties and architectural design that have become more accessible and innovative, enhancing appeal to prospective buyers and opening new design possibilities for architects.

Within the financial sector, GANs are leveraged to simulate fraudulent activities, thus contributing to robust fraud detection systems. Investment firms are also exploring GANs to model complex market scenarios, shaping optimized investment strategies. Lastly, the environmental and urban planning fields have found unique applications for GANs. Whether modeling complex climate patterns to inform policy or visualizing urban growth for sustainable development, GANs offer unprecedented capabilities. The widespread adoption of GANs across these sectors demonstrates their multifaceted nature. Their generative power is not only enhancing efficiency and reducing costs but also forging new paths in creativity, personalization, and predictive insight. As GAN technology continues to evolve, it is poised to redefine traditional business practices further, aligning technology and innovation in ways that were previously unimaginable.

CONCLUSIONS

We observed how GANs and VAEs, while primarily rooted in the domain of generative modeling, offer insights into data replication and creation, challenging our perceptions of reality in digital realms. The transformer architecture has shown how profound understanding can be garnered from data sequences, leading to innovations that seemed distant a few decades ago.

Moreover, the dance of decision-making showcased by RL elucidated the importance of interaction and feedback in learning processes. The exploration of its intersections with the aforementioned models highlighted the burgeoning potential of combining representational strength with decision-making acumen.

It's undeniable that each of these models and architectures holds its significance in isolation. However, their true potential, as our exploration suggests, may lie in their collaborative application. The melding of generative capabilities with sequential processing and decision-making paradigms is not merely an academic pursuit. It signifies a direction for the future, where boundaries blur, interdisciplinary engagements flourish, and the convergence of varied methodologies will hopefully lead to solutions for challenges considered insurmountable.

REFERENCES

Abraham, A. (2005). Artificial neural networks. *Handbook of Measuring System Design.* http://wsc10.softcomputing.net/ann_chapter.pdf

Ahmed, S., & Chua, H. W. (2023). Perception and deception: Exploring individual responses to deepfakes across different modalities. *Heliyon, 9*(10), e20383.

Akiti, C. (2021). *Efficient Few-Shot Learning for Named Entity Recognition.* Pennsylvania State University.

Alekseev, A., & Bobe, A. (2019). GaborNet: Gabor filters with learnable parameters in deep convolutional neural networks. In *arXiv [cs.CV].* arXiv. https://ieeexplore.ieee.org/abstract/document/9030571/

Aljarah, I., Faris, H., & Mirjalili, S. (2018). Optimizing connection weights in neural networks using the whale optimization algorithm. *Soft Computing*, *22*(1), 1–15.

Alpaydin, E. (2021). *Machine Learning, Revised and Updated Edition*. MIT Press.

Arel, I. (2012). Deep reinforcement learning as foundation for artificial general intelligence. In P. Wang & B. Goertzel (Eds.), *Theoretical Foundations of Artificial General Intelligence* (pp. 89–102). Atlantis Press.

Arjovsky, M., Chintala, S., & Bottou, L. (06–11 August 2017). Wasserstein Generative adversarial networks. In D. Precup & Y. W. Teh (Eds.), *Proceedings of the 34th International Conference on Machine Learning* (Vol. 70, pp. 214–223). PMLR.

Atienza, R. (2020). *Advanced Deep Learning with TensorFlow 2 and Keras: Apply DL, GANs, VAEs, deep RL, Unsupervised Learning, Object Detection and Segmentation, and more, 2nd Edition*. Packt Publishing Ltd.

Ba, J. L., Kiros, J. R., & Hinton, G. E. (2016). Layer normalization. In *arXiv [stat.ML]*. arXiv. http://arxiv.org/abs/1607.06450

Babcock, J., & Bali, R. (2021). *Generative AI with Python and TensorFlow 2: Create Images, Text, and Music with VAEs, GANs, LSTMs, Transformer Models*. Packt Publishing Ltd.

Bhargava, C., & Sharma, P. K. (2021). *Artificial Intelligence: Fundamentals and Applications*. CRC Press.

Bond-Taylor, S., Leach, A., Long, Y., & Willcocks, C. G. (2022). Deep generative modelling: A comparative review of VAEs, GANs, normalizing flows, energy-based and autoregressive models. *IEEE Transactions on Pattern Analysis and Machine Intelligence*, *44*(11), 7327–7347.

Bory, P. (2019). Deep new: The shifting narratives of artificial intelligence from Deep Blue to AlphaGo. *Convergence*, *25*(4), 627–642.

Carpenter, G. A. (1989). Neural network models for pattern recognition and associative memory. *Neural Networks: The Official Journal of the International Neural Network Society*, *2*(4), 243–257.

Cassano, F., Gouwar, J., Lucchetti, F., Schlesinger, C., Anderson, C. J., Greenberg, M., Jangda, A., & Guha, A. (2023). Knowledge transfer from high-resource to low-resource programming languages for code LLMs. In *arXiv [cs.PL]*. arXiv. http://arxiv.org/abs/2308.09895

Chai, J. (2022). *Graph-based Methods for Zero-shot Learning in Image Classifications*. Pennsylvania State University.

Chen, X., Han, Y., & Zhang, J. (2023). APRIL-GAN: A zero-/few-shot anomaly classification and segmentation method for CVPR 2023 VAND workshop challenge tracks 1&2: 1st place on zero-shot AD and 4th place on few-shot AD. In *arXiv [cs.CV]*. arXiv. http://arxiv.org/abs/2305.17382

Coronato, A., Naeem, M., De Pietro, G., & Paragliola, G. (2020). Reinforcement learning for intelligent healthcare applications: A survey. *Artificial Intelligence in Medicine*, *109*, 101964.

Cui, B., Li, Y., Chen, M., & Zhang, Z. (2019). Fine-tune BERT with sparse self-attention mechanism. *Proceedings of the 2019 Conference on Empirical Methods in Natural Language Processing and the 9th International Joint Conference on Natural Language Processing (EMNLP-IJCNLP)*, 3548–3553.

Cunningham, P., Cord, M., & Delany, S. J. (2008). Supervised learning. In M. Cord & P. Cunningham (Eds.), *Machine Learning Techniques for Multimedia: Case Studies on Organization and Retrieval* (pp. 21–49). Springer Berlin Heidelberg.

da Silva, I. N., Spatti, D. H., Flauzino, R. A., Liboni, L. H. B., & dos Reis Alves, S. F. (2016). *Artificial Neural Networks: A Practical Course*. Springer.

Dagan, G., Keller, F., & Lascarides, A. (2023). Dynamic planning with a LLM. In *arXiv [cs.CL]*. arXiv. http://arxiv.org/abs/2308.06391

Devlin, J., Chang, M.-W., Lee, K., & Toutanova, K. (2018). BERT: Pre-training of deep bidirectional transformers for language understanding. In *arXiv [cs.CL]*. arXiv. http://arxiv.org/abs/1810.04805

Dobrin, S. I. (2023). *AI and Writing*. Broadview Press.

Doersch, C. (2016). Tutorial on variational autoencoders. In *arXiv [stat.ML]*. arXiv. http://arxiv.org/abs/1606.05908

Du, Z., Qian, Y., Liu, X., Ding, M., Qiu, J., Yang, Z., & Tang, J. (2021). All nlp tasks are generation tasks: A general pretraining framework. In *arXiv preprint arXiv:2103. 10360* (Vol. 18). Mar. http://keg.cs.tsinghua.edu.cn/jietang/publications/ACL22-Du-et-al-GLM.pdf

Durall, R., Frolov, S., Hees, J., Raue, F., Pfreundt, F.-J., Dengel, A., & Keupe, J. (2021). Combining transformer generators with convolutional discriminators. In *arXiv [cs.CV]*. arXiv. http://arxiv.org/abs/2105.10189

El Naqa, I., & Murphy, M. J. (2015). What is machine learning? In I. El Naqa, R. Li, & M. J. Murphy (Eds.), *Machine Learning in Radiation Oncology: Theory and Applications* (pp. 3–11). Springer International Publishing.

El Sallab, A., Abdou, M., Perot, E., & Yogamani, S. (2017). Deep reinforcement learning framework for autonomous driving. In *arXiv [stat.ML]*. arXiv. http://arxiv.org/abs/1704.02532

Finnie-Ansley, J., Denny, P., Luxton-Reilly, A., Santos, E. A., Prather, J., & Becker, B. A. (2023, January 30). My AI wants to know if this will be on the exam: Testing OpenAI's codex on CS2 programming exercises. *Australasian Computing Education Conference*. ACE '23: Australasian Computing Education Conference, Melbourne VIC Australia. https://doi.org/10.1145/3576123.3576134

Firc, A., Malinka, K., & Hanáček, P. (2023). Deepfakes as a threat to a speaker and facial recognition: An overview of tools and attack vectors. *Heliyon, 9*(4), e15090.

Floridi, L., & Chiriatti, M. (2020). GPT-3: Its nature, scope, limits, and consequences. *Minds and Machines, 30*(4), 681–694.

Font, J. M., & Mahlmann, T. (2021). The Dota 2 Bot competition. In *arXiv [cs.AI]*. arXiv. https://ieeexplore.ieee.org/abstract/document/8356682/

Fröhling, L., & Zubiaga, A. (2021). Feature-based detection of automated language models: Tackling GPT-2, GPT-3 and Grover. *PeerJ. Computer Science, 7*, e443.

Geenjaar, E., Lewis, N., Kashyap, A., Miller, R., & Calhoun, V. (2022). CommsVAE: Learning the brain's macroscale communication dynamics using coupled sequential VAEs. In *arXiv [q-bio.NC]*. arXiv. http://arxiv.org/abs/2210.03667

Goodfellow, I., Pouget-Abadie, J., Mirza, M., Xu, B., Warde-Farley, D., Ozair, S., Courville, A., & Bengio, Y. (2014). Generative adversarial nets. *Advances in Neural Information Processing Systems, 27*. https://proceedings.neurips.cc/paper_files/paper/2014/hash/5ca3e9b122f61f8f06494c97b1afccf3-Abstract.html

Gozalo-Brizuela, R., & Garrido-Merchan, E. C. (2023). ChatGPT is not all you need. A state of the art review of large generative AI models. In *arXiv [cs.LG]*. arXiv. http://arxiv.org/abs/2301.04655

Gulrajani, I., Ahmed, F., Arjovsky, M., Dumoulin, V., & Courville, A. C. (2017). Improved training of wasserstein gans. *Advances in Neural Information Processing Systems, 30*. https://proceedings.neurips.cc/paper_files/paper/2017/hash/892c3b1c6dccd52936e27cbd0ff683d6-Abstract.html

Hall, J. R., & Sparks, T. D. (2023). A case study of beta-variational auto-encoders, disentanglement impacts of input distribution and beta-variation based upon a computational multi-modal particle packing simulation. *Integrating Materials and Manufacturing Innovation*. https://doi.org/10.1007/s40192-023-00306-6

Hendry, M. F., Kottmann, N., Fröhlich, M., Bruggisser, F., Quandt, M., Speziali, S., Huber, V., & Salter, C. (2023). Are you talking to me? A case study in emotional human-machine interaction. *Proceedings of the AAAI Conference on Artificial Intelligence and Interactive Digital Entertainment, 19*(1), 417–424.

Hochreiter, S., & Schmidhuber, J. (1997). Long short-term memory. *Neural Computation*, *9*(8), 1735–1780.

Hogan, T. (2023). *The Transformer Architecture: A Practical Guide to Natural Language Processing*. Amazon Digital Services LLC – Kdp.

Holcomb, S. D., Porter, W. K., Ault, S. V., Mao, G., & Wang, J. (2018). Overview on DeepMind and its AlphaGo Zero AI. *Proceedings of the 2018 International Conference on Big Data and Education*, 67–71.

Hopfield, J. J. (1988). Artificial neural networks. *IEEE Circuits and Devices Magazine*, *4*(5), 3–10.

Huang, H., Feng, Y., Shi, C., Xu, L., Yu, J., & Yang, S. (2023). Free-bloom: Zero-shot Text-to-video generator with LLM Director and LDM Animator. In *arXiv [cs.CV]*. arXiv. http://arxiv.org/abs/2309.14494

Huck, M. (2023). *GPT-4 Complete: A Comprehensive Technical Guide to the New OpenAI Model*. Amazon Digital Services LLC – Kdp.

Iliadis, L., Papaleonidas, A., Angelov, P., & Jayne, C. (2023). *Artificial Neural Networks and Machine Learning – ICANN 2023: 32nd International Conference on Artificial Neural Networks, Heraklion, Crete, Greece, September 26–29, 2023, Proceedings, Part V*. Springer Nature.

Isola, P., Zhu, J.-Y., Zhou, T., & Efros, A. A. (2016). Image-to-image translation with conditional adversarial networks. In *arXiv [cs.CV]* (pp. 1125–1134). arXiv. http://openaccess.thecvf.com/content_cvpr_2017/html/Isola_Image-To-Image_Translation_With_CVPR_2017_paper.html

Kampik, T., Warmuth, C., Rebmann, A., Agam, R., Egger, L. N. P., Gerber, A., Hoffart, J., Kolk, J., Herzig, P., Decker, G., van der Aa, H., Polyvyanyy, A., Rinderle-Ma, S., Weber, I., & Weidlich, M. (2023). Large process models: Business process management in the age of generative AI. In *arXiv [cs.SE]*. arXiv. http://arxiv.org/abs/2309.00900

Katz, D. M., Bommarito, M. J., Gao, S., & Arredondo, P. (2023). *GPT-4 Passes the Bar Exam*. https://doi.org/10.2139/ssrn.4389233

King, M. (2023). Can GPT-4 formulate and test a novel hypothesis? Yes and no. In *TechRxiv*. https://doi.org/10.36227/techrxiv.22517278.v1

Kingma, D. P., & Welling, M. (2019). *An Introduction to Variational Autoencoders*. Now Publishers.

Koubaa, A. (2023). GPT-4 vs. GPT-3.5: A concise showdown. In *Preprints*. https://doi.org/10.20944/preprints202303.0422.v1

Lee, G., Kim, H., Kim, J., Kim, S., Ha, J.-W., & Choi, Y. (2022). Generator knows what discriminator should learn in unconditional GANs. *Computer Vision – ECCV 2022*, 406–422. https://arxiv.org/abs/2207.13320

Li, X., Yao, Y., Jiang, X., Fang, X., Meng, X., Fan, S., Han, P., Li, J., Du, L., Qin, B., Zhang, Z., Sun, A., & Wang, Y. (2023). FLM-101B: An open LLM and how to train it with $100K Budget. In *arXiv [cs.CL]*. arXiv. http://arxiv.org/abs/2309.03852

Li, Y. (2017). Deep reinforcement learning: An overview. In *arXiv [cs.LG]*. arXiv. http://arxiv.org/abs/1701.07274

Liu, X., & Hsieh, C.-J. (2019). Rob-gan: Generator, discriminator, and adversarial attacker. *Proceedings of the IEEE/CVF Conference on Computer Vision and Pattern Recognition*, 11234–11243.

Madaan, A., Tandon, N., Gupta, P., Hallinan, S., Gao, L., Wiegreffe, S., Alon, U., Dziri, N., Prabhumoye, S., Yang, Y., Gupta, S., Majumder, B. P., Hermann, K., Welleck, S., Yazdanbakhsh, A., & Clark, P. (2023). Self-refine: Iterative refinement with self-feedback. In *arXiv [cs.CL]*. arXiv. http://arxiv.org/abs/2303.17651

Manning, C. D., Surdeanu, M., Bauer, J., Finkel, J. R., Bethard, S., & McClosky, D. (2014). The Stanford CoreNLP natural language processing toolkit. *Proceedings of 52nd Annual Meeting of the Association for Computational Linguistics: System Demonstrations*, 55–60.

Martin, J. T. (2023). Hello, LaMDA! *Brain: A Journal of Neurology, 146*(3), 793–795.

Meikle, G. (2022). *Deepfakes*. John Wiley & Sons.

Miladinović, Đ., Shridhar, K., Jain, K., Paulus, M. B., Buhmann, J. M., Sachan, M., & Allen, C. (2022). Learning to drop out: An adversarial approach to training sequence VAEs. In S. Koyejo, S. Mohamed, A. Agarwal, D. Belgrave, K. Cho, & A. Oh (Eds.), *arXiv [cs. LG]* (pp. 9645–9659). arXiv. https://proceedings.neurips.cc/paper_files/paper/2022/file/3ed57b293db0aab7cc30c44f45262348-Paper-Conference.pdf

Mirza, M., & Osindero, S. (2014). Conditional generative adversarial nets. In *arXiv [cs.LG]*. arXiv. http://arxiv.org/abs/1411.1784

Mnih, V., Kavukcuoglu, K., Silver, D., Graves, A., Antonoglou, I., Wierstra, D., & Riedmiller, M. (2013). Playing Atari with deep reinforcement learning. In *arXiv [cs.LG]*. arXiv. http://arxiv.org/abs/1312.5602

Nasteski, V. (2017). An overview of the supervised machine learning methods. *Horizons. B, 4*, 51–62.

Ngiam, J., Khosla, A., Kim, M., Nam, J., Lee, H., & Ng, A. Y. (2011). *Multimodal Deep Learning*. https://openreview.net/pdf?id=Hk4OO3W_bS

Ozdemir, S. (2023). *Quick Start Guide to Large Language Models: Strategies and Best Practices for Using ChatGPT and Other LLMs*. Pearson.

Peng, H., Schwartz, R., & Smith, N. A. (2019). PaLM: A hybrid parser and language model. In *arXiv [cs.CL]*. arXiv. http://arxiv.org/abs/1909.02134

Peng, L., Zhang, Y., & Shang, J. (2023). Generating efficient training data via LLM-based attribute manipulation. In *arXiv [cs.CL]*. arXiv. http://arxiv.org/abs/2307.07099

Poels, Y., & Menkovski, V. (2022). VAE-CE: Visual contrastive explanation using disentangled VAEs. *Advances in Intelligent Data Analysis XX*, 237–250.

Polydoros, A. S., & Nalpantidis, L. (2017). Survey of model-based reinforcement learning: Applications on robotics. *Journal of Intelligent and Robotic Systems, 86*(2), 153–173.

Radford, A., Metz, L., & Chintala, S. (2015). Unsupervised representation learning with deep convolutional generative adversarial networks. In *arXiv [cs.LG]*. arXiv. http://arxiv.org/abs/1511.06434

Salehinejad, H., Sankar, S., Barfett, J., Colak, E., & Valaee, S. (2017). Recent advances in recurrent neural networks. In *arXiv [cs.NE]*. arXiv. http://arxiv.org/abs/1801.01078

Sanderson, K. (2023). GPT-4 is here: What scientists think. *Nature, 615*(7954), 773.

Sarkar, A., & Cooper, S. (2021). Dungeon and platformer level blending and generation using conditional VAEs. In *arXiv [cs.LG]*. arXiv. http://arxiv.org/abs/2106.12692

Sewak, M. (n.d.). *Deep Reinforcement Learning*. Springer Nature Singapore.

Stokel-Walker, C., & Van Noorden, R. (2023). What ChatGPT and generative AI mean for science. *Nature, 614*(7947), 214–216.

Sutton, R. S., & Barto, A. G. (2018). *Reinforcement Learning, Second Edition: An Introduction*. MIT Press.

Tato, A., & Nkambou, R. (2018). *Improving Adam optimizer*. https://openreview.net/pdf?id=HJfpZq1DM

Tian, J., Yan, B., Yu, J., Weng, C., Yu, D., & Watanabe, S. (2022). Bayes risk CTC: Controllable CTC alignment in Sequence-to-Sequence tasks. In *arXiv [cs.CL]*. arXiv. http://arxiv.org/abs/2210.07499

Unsupervised Learning Algorithms. (n.d.). Springer International Publishing.

Wan, X., Sun, R., Dai, H., Arik, S. O., & Pfister, T. (2023). Better zero-shot reasoning with self-adaptive prompting. In *arXiv [cs.CL]*. arXiv. http://arxiv.org/abs/2305.14106

Wang, P., Panda, R., Hennigen, L. T., Greengard, P., Karlinsky, L., Feris, R., Cox, D. D., Wang, Z., & Kim, Y. (2023). Learning to grow pretrained models for efficient transformer training. In *arXiv [cs.LG]*. arXiv. http://arxiv.org/abs/2303.00980

Wang, Z., Yang, Z., Song, X., Zhang, H., Sun, B., Zhai, J., Yang, S., Xie, Y., & Liang, P. (2024). Raman spectrum model transfer method based on Cycle-GAN. *Spectrochimica Acta. Part A, Molecular and Biomolecular Spectroscopy, 304*, 123416.

Wei, J., Wang, X., Schuurmans, D., Bosma, M., Ichter, B., Xia, F., Chi, E., Le, Q., & Zhou, D. (2022). Chain-of-thought prompting elicits reasoning in large language models. In S. Koyejo, S. Mohamed, A. Agarwal, D. Belgrave, K. Cho, & A. Oh (Eds.), *arXiv [cs.CL]* (pp. 24824–24837). arXiv. https://proceedings.neurips.cc/paper_files/paper/2022/file/9d5609613524ecf4f15af0f7b31abca4-Paper-Conference.pdf

Wu, Y., Wang, S., Song, G., & Huang, Q. (2019). Learning fragment self-attention embeddings for image-text matching. *Proceedings of the 27th ACM International Conference on Multimedia*, 2088–2096.

Yao, K., Peng, B., Zhang, Y., Yu, D., Zweig, G., & Shi, Y. (2014). Spoken language understanding using long short-term memory neural networks. *2014 IEEE Spoken Language Technology Workshop (SLT)*, 189–194.

Ying, X. (2019). An overview of overfitting and its solutions. *Journal of Physics. Conference Series, 1168*(2), 022022.

Yu, Y., Zhuang, Y., Zhang, J., Meng, Y., Ratner, A., Krishna, R., Shen, J., & Zhang, C. (2023). Large language model as attributed training data generator: A tale of diversity and bias. In *arXiv [cs.CL]*. arXiv. http://arxiv.org/abs/2306.15895

Zhang, H., Xu, T., Li, H., Zhang, S., Wang, X., Huang, X., & Metaxas, D. (2017). StackGAN: Text to photo-realistic image synthesis with stacked generative adversarial networks. *2017 IEEE International Conference on Computer Vision (ICCV)*, 5907–5915.

Zhang, Z. (2018). Improved adam optimizer for deep neural networks. *2018 IEEE/ACM 26th International Symposium on Quality of Service (IWQoS)*, 1–2.

Zhu, J.-Y., Zhang, R., Pathak, D., Darrell, T., Efros, A. A., Wang, O., & Shechtman, E. (2017). Toward multimodal image-to-image translation. *Advances in Neural Information Processing Systems, 30*. https://proceedings.neurips.cc/paper_files/paper/2017/hash/819f46e52c25763a55cc642422644317-Abstract.html

3 Collaborative AI and the Future of Work

INTRODUCTION

The theme of collaboration is hardly new; it has been the backbone of human innovation and progress. Historically, collaboration has been a defining feature of human advancement. Whether it is the collective efforts that led to the construction of the ancient pyramids, the intellectual synergy among scientists that sparked the Renaissance, or the cross-border cooperation that enabled space exploration, collaboration has always been the catalyst for reaching new frontiers (Malone & Bernstein, 2022).

In business, politics, science, and art, collaboration has allowed individuals to pool their diverse skills, knowledge, and perspectives to achieve goals that would have been unreachable by a single individual. Through collaboration the barriers have been broken, limitations overcome, and new horizons discovered.

The emergence of AI presents a novel dimension to the age-old concept of collaboration. In the labyrinth of modern work environments, the idea of collaboration often emerges as the holy grail of productivity and innovation (Cross, 2021; Autor, 2015; Acemoglu & Restrepo, 2017a; Acemoglu & Restrepo, 2018; Acemoglu et al., 2019). Like in an orchestra, every musician attuned not just to their own instrument but to the entire ensemble, creating a symphony greater than the sum of its parts. Now replace a few musicians with advanced algorithms or robotic arms. Does the music still resonate with the same harmony? Welcome to the evolving landscape of human-AI collaboration, where traditional ensembles are being disrupted and redefined.

Traditional human collaboration relies on a multitude of verbal, nonverbal, and subliminal cues (Nelson & Lundin, 2010). We have developed elaborate social contracts and ethical frameworks to guide our interactions based on the tacit knowledge we accumulated over a lifetime of trial and error of communicating with other humans.

Version 1.0 of integrating nonhuman agents into the work ecosystem can be defined by human-machine collaboration. Assembly line robots or a programmable coffee maker performs predefined tasks, often more efficiently and tirelessly than a human could. But they lack the ability to adapt and understand the nuances of human behavior and decision-making. The interaction is strictly one-sided: you press a button; the machine springs into action. This collaboration is effective but not exactly a partnership.

Enter human-AI collaboration, version 2.0 if you will, has a twist that is more revolutionary rather than iterative. Modern AI systems bring something new to the

DOI: 10.1201/9781032656618-4

table: the ability to learn, adapt, and sometimes even "understand" human behavior to a certain extent. They process large datasets, identify patterns, and make recommendations based on complex algorithms. Unlike their traditional machine predecessors, AI systems can anticipate needs, suggest optimizations, and flag ethical or logical inconsistencies (Feldman, 2017).

In this updated framework, AI functions as a collaborative partner that can significantly extend human cognitive capabilities. AI systems allow for faster data processing and automation (Harvard Business Review et al., 2019) and include a deeper, more integrated form of interaction that amplifies human decision-making, analysis, and creativity. AI becomes more than a tool; it becomes a partner, a collaborator, and, in some instances, a muse. It offers the possibility of augmenting human abilities, providing insights based on massive data sets, executing complex computations in seconds, and generating creative content that resonates with human emotions.

This symbiosis gives rise to what we refer to as modern human unicorns – individuals who are empowered by AI to reach unprecedented levels of skill, creativity, and innovation. These are not mythical creatures or innovative startup companies but real professionals whose abilities are amplified by AI.

In this chapter, we will explore the dynamics of human-AI partnership, focusing on how collaborative AI is shaping the future of diverse fields including augmenting human capabilities with generative AI, new roles and skills in a collaborative AI ecosystem, and the impact of collaborative AI on employment and the labor market.

AUGMENTING HUMAN CAPABILITIES WITH GENERATIVE AI: HUMAN UNICORN

The concept of a unicorn has its roots in the dawn of humanity. In the world of myths and legends, the unicorn is a symbol of rarity and uniqueness. It is a creature so elusive that its very existence is the subject of debate. With its singular horn and often-depicted magical abilities, the unicorn captures our imagination precisely because it defies common experience.

A metaphor of unicorn has been used in business and technology to describe an entity that is not just incrementally better but qualitatively different. A metaphor of a human unicorn may be used to describe a human who possesses extraordinary abilities challenging the boundaries of what we think is possible.

In ancient times, a human unicorn might have been a visionary leader, a masterful artist, or a pioneering scientist. These were individuals who stood out for their exceptional talents, wisdom, or insights, leading their societies to new frontiers of understanding and achievement.

One exemplary human unicorn that embodies this rare combination of abilities is Leonardo da Vinci, a polymath of the Renaissance era. Da Vinci's work transcended traditional boundaries, as he excelled in various fields including art, science, engineering, anatomy, and more. His curiosity and insatiable thirst for knowledge drove him to explore a vast array of subjects, each with a depth and precision that were truly remarkable for his time.

As an artist, da Vinci created some of the most iconic works in art history, such as the *Mona Lisa* and *The Last Supper.* His meticulous attention to detail, understanding of human anatomy, and ability to capture emotion set a new standard for artistic expression. But his talents did not stop there. Da Vinci was equally fascinated by the natural world and the mechanics of the human body. His study of human anatomy resulted in detailed sketches and observations that contributed to the field of medicine. His anatomical drawings were not only artistically profound but scientifically accurate, providing insights that were ahead of his time.

In engineering, da Vinci's imaginative mind conjured inventions that were precursors to modern machines, such as helicopters, tanks, and submarines. Though many of his inventions were never built in his lifetime, his sketches and notes demonstrate a deep understanding of mechanical principles and an ability to envision complex machinery.

In many ways, Leonardo da Vinci's life and work serve as a timeless example of what it means to be a human unicorn. He was not confined to a single domain but rather thrived in multiple disciplines, integrating art with science, observation with imagination, and creativity with logic. His ability to see connections between disparate fields and to innovate across them made him a true unicorn of his era. His legacy continues to inspire modern thinkers, innovators, and creators who strive to transcend traditional boundaries and explore the interconnectedness of knowledge and creativity. In fact, the best performing generative pretrained transformer model by OpenAI is named after da Vinci genius, encapsulating the essence of his multidisciplinary excellence, innovation, precision, pioneering spirit, enduring influence, and a humanistic approach.

Today's human unicorn leverages generative AI to synthesize vast amounts of information, using algorithms to create works that transcend traditional boundaries. The AI acts as an extension of human creativity, fostering innovation across various domains. While da Vinci's talents were constrained by the technologies and knowledge of his time, the modern unicorn has the advantage of AI to analyze and integrate diverse sources of information, leading to broader and more nuanced inventions.

In creative fields, such as writing, AI's capacity to sift through vast literary landscapes means access to a plethora of styles, genres, and cultural insights, enabling the creation of works that are both original and reflective of the author's perspectives. A musician utilizing AI can blend the mathematical precision of algorithms with the emotional resonance of human composition, crafting melodies that transcend traditional musical boundaries. In visual arts, the collaboration with AI opens doors to experiment with an infinite array of textures, colors, and techniques, creating paintings that marry centuries-old traditions with cutting-edge innovations.

Compared with the modern unicorn, leveraging the power of AI, is similar to da Vinci's journey but with tools that amplify and augment human ability. While da Vinci meticulously studied the human form and translated it into art and scientific knowledge, the modern unicorn engages in a partnership with AI to push the boundaries of creation and exploration. For example, in healthcare data analytics, AI can process data to detect early signs of diseases like cancer. A recent study in Sweden (Lång et al., 2023) evaluated the results of more than 80,000 mammograms and concluded that AI-supported mammogram screening increases breast cancer detection by 20% augmenting the radiologist's abilities to diagnose abnormal cell growth early and at the same time reducing the burden of the excessive amount of reading.

The machine's ability to discern minute abnormalities, which might escape the human eye, is an extension of da Vinci's own dedication to understanding the intricacies of the human body. Just as da Vinci's sketches revealed the hidden structures of the human form, AI unveils insights hidden in complex data.

Da Vinci's technological visions were grand but often unrealized, restricted to the pages of his notebooks. He designed groundbreaking concepts for flying machines, mechanical devices, and innovative bridges, but they largely remained theoretical, held back by the technological constraints of his era. The modern unicorn, by contrast, taps into the power of AI to transform ideas into tangible creations. In today's world, AI's capabilities extend into complex simulations and precise modeling across various engineering workflows. Let's consider car safety research. According to the World Health Organization (2023), each year 1.19 million people die as a result of road traffic crashes, and between 20 and 50 million more people suffer nonfatal injuries, with many incurring a disability as a result. To improve car safety, engineers collaborate with AI to analyze an array of variables such as speed, impact angle, weather, and road conditions, to understand the effects on crash outcomes. This collaboration leads to vehicles designed with the utmost safety considerations in mind. Companies like Monolith AI, partnering with leading automobile manufacturers like BMW, utilize machine sensor data to create precise, AI-powered, self-learning models to increase the safety of vehicles (Monolith AI, 2023). These models quickly grasp and predict the outcomes of complex physics, a process that may have once seemed as fantastical as da Vinci's flying machines.

In engineering, human unicorns also turn imagination into reality. From smart transportation systems to renewable energy solutions, AI-driven algorithms help engineers model, simulate, and optimize complex systems. For example, in the design of earthquake-resistant buildings, modern engineers utilize AI to simulate various seismic scenarios, ensuring that structures can withstand conditions of great magnitude (Er Akan et al., 2023). The synergy between human and machine is also evident in the field of software development where AI proactively detects bugs and suggests improvements. This is a stark contrast to da Vinci's era, where the realization of intricate designs depended solely on human ingenuity. Today, the software developer's creativity is amplified by AI, allowing for more sophisticated and responsive solutions.

In the medical field, AI-assisted robotic surgery enables surgeons to perform complex procedures in remote places with greater precision, flexibility, and control. The surgery is not actually performed by robots. Instead, the surgeon translates their hand movements into smaller, more precise movements of tiny instruments inside the patient's body, augmenting the surgeon's ability to perform precise procedures remotely with the assistance of AI. In fact, one of the well-known surgical robots, designed at the University of California Berkeley, is named after da Vinci (Metz, 2021). Generative AI can augment this process further by allowing voice commands to interact with the system, providing real-time analysis and feedback, and even predicting potential complications based on the patient's unique medical history. These advancements represent a transformative shift in the medical landscape, where the synergy between human intuition and machine intelligence brings a new level of efficiency, accuracy, and safety to surgical practices. By harnessing the power of AI, the modern surgical environment evolves into a dynamic ecosystem where human expertise is elevated and complemented by the sophistication of artificial intelligence, redefining what's possible in healthcare and reflecting the broader trend of human-AI collaboration.

In pharmaceuticals, AI algorithms speed up drug discovery, predicting interactions and side effects, and creating revolutionary new proteins that can be used for creating novel medications and therapies to combat diseases, engineering resilient crops for changing climates, and developing enzymes capable of degrading pervasive environmental plastics (Callaway, 2022). These capabilities showcase how AI transcends the automation of tasks and becomes a creative and collaborative partner in complex problem-solving. By working in tandem with human scientists, AI helps navigate intricate biological systems, uncovering novel insights that might take years to discover through traditional means. In turn, these advancements expedite the development and delivery of life-saving drugs, promote sustainable agriculture, and contribute to environmental preservation (Minati & Pessa, 2012).

In manufacturing, human-AI collaboration leads to the creation of tailored products that reflect the unique needs and desires of the consumer. Through the utilization of AI-driven 3D printing technologies, production of otherwise unfeasible or labor-intensive structures becomes not only possible but efficient. A notable example is Nvidia's Magic3D, a generative AI that turns text prompts into intricate 3D models, aiding in the visualization and conceptualization of components with multifaceted and organic forms (Lin et al., 2023). The connection between generative AI for 3D modeling and 3D printing technology manifests a synergy that fosters innovation and cost-effectiveness and revolutionizes the way we design and produce, taking manufacturing into a new era where human imagination is augmented by AI's capability. This transformation plays a pivotal role in democratizing the development of new products. By leveraging AI's computational power, small businesses and independent creators can access tools and technologies that were previously restricted to large corporations and research institutions. This accessibility bridges the gap between idea and realization, allowing more individuals to contribute to the ever-evolving landscape of innovation. The creative designs that remained confined to da Vinci's notebooks can now be brought to life, ranging from the form of intricate jewelry to aerospace components.

Across other scientific disciplines, from astronomy to environmental science, generative AI has emerged as a collaborative partner, fostering innovation. In space exploration, the National Aeronautics and Space Administration (NASA) has employed generative AI to create elaborate parts for new spaceships, leading to efficiencies in design and manufacturing that were previously unattainable (Lal et al., 2022). In astrophysics, generative AI plays a vital role in the analysis of immense and complex data generated by telescopes. It is employed for computational tasks and to actively generate new models and insights into the universe's formation, structure, and the very fabric of space-time. By generating hypotheses and deciphering patterns that might otherwise elude human observation, generative AI contributes to a deeper understanding of cosmic phenomena.

Environmental science is benefiting from the use of generative AI, particularly in the area of real-time monitoring and prediction of natural disasters. Generative AI can create complex simulations based on vast quantities of data from weather stations, satellite imagery, and seismic sensors. These simulations allow for more accurate and timely predictions, leading to better preparedness and more effective response strategies. By generating predictive models that are fine-tuned to local conditions, generative AI has the potential to save lives. Its role in environmental monitoring extends to generating new insights into climate change, ecological balance, and crafting innovative solutions to pressing environmental issues.

WHAT DOES IT MEAN TO BE A MODERN UNICORN?

To answer this question, we will refer to ChatGPT to provide a definition of a modern unicorn. As unicorns may reside in different realms, we will explore the definitions from various perspectives using different prompts. By employing humorous, academic, and marketing prompts, we will showcase how the language model's response shifts, altering the tone, context, and insights. This exercise will provide a multifaceted view of the human unicorn concept and demonstrate the remarkable adaptability and responsiveness of generative AI to produce diverse content. By exploring the same subject from various angles, we invite you on a unique intellectual journey of exploring modern unicorns.

Creating effective prompts for generative AI is an art form in itself, and collecting these valuable prompts into a personalized library can serve as a powerful resource for future use, enabling more efficient and targeted interactions with the technology.

THE POWER OF PROMPTS

Use Case 1 – Add Some Humor

Prompt: Write a humorous article titled "A Day in the Life of a Human Unicorn." Keep the tone light, whimsical, and filled with humor.

A Day in the Life of a Human Unicorn

If you've ever seen a unicorn, chances are it was on a child's backpack, a popular logo, or in your dreams after consuming too much spicy food late at night. But have you ever met a human unicorn? You know, those rare individuals who combine the magical powers of generative AI with their extraordinary human brains to prance through life's challenges, sprinkling glitter wherever they go? Allow me to introduce you to a day in the life of one such specimen.

6:00 AM – The Morning Glitter Routine: A human unicorn's day starts not with a shrill alarm but a chorus of virtual nightingales serenading them awake. They then check their AI-powered "glitter forecast," which predicts the optimal amount of sparkle needed for the day's activities.

Breakfast, of course, is a bowl of rainbow cereal, eaten with a golden spoon that plays motivational ballads as it approaches the mouth. It's an AI invention, after all, and it knows what's needed to start the day right!

9:00 AM – A Prance into the Office: Why walk when you can prance? With their augmented reality glasses showing a trail of sparkles, the human unicorn commutes to work. Along the way, they help a lost kitten, find a lost love, and draft three business proposals – all with the help of generative AI and a swish of their imaginary tail.

11:00 AM – The Magical Meeting Room: Imagine a conference room filled with ordinary folks yawning through a PowerPoint. Enter the human unicorn! Suddenly, graphs come to life, charts dance, and a holographic rainbow arcs over the room. Decision-making is done in a blink, thanks to AI-driven insights, leaving more time for team-building activities like trust falls into virtual clouds.

2:00 PM – Lunch: A Whimsical Feast: Lunch is no mere meal for a human unicorn; it's a gastronomic adventure! With the help of AI, they conjure a menu that not only satisfies hunger but inspires creativity, unity, and an unexplained urge to hum show tunes.

3:00 PM – Solving World Problems, One Sparkle at a Time: Post-lunch, the human unicorn tackles real-world problems. From peace negotiations conducted on virtual reality bridges to sustainable solutions wrapped in digital silk ribbons, nothing is beyond their reach.

6:00 PM – Home Again to the Unicorn Lair: A human unicorn's home is a place of wonder. From the AI-powered self-cleaning glitter floor to the enchanted smart fridge that sings recipes, it's a marvel of modern technology and mythical charm.

9:00 PM – Bedtime, But Not Before…: Before settling into their cloudlike bed, the human unicorn checks their "generative dreamscape," setting up scenarios for the night's dreams. Pirates? Spaceships? A combination of both? The choices are endless!

Conclusion: The life of a human unicorn is not merely a life; it's a kaleidoscope of imagination, ingenuity, and a hint of absurdity. They teach us that with a dash of magic (and a sprinkle of AI), we can all find the unicorn within.

So next time you're stuck in traffic, just remember: Somewhere out there, a human unicorn is probably prancing past cars, with generative AI guiding the way, and a sparkle in their eye. Why not join them? You might just find that life's a little more magical when you're a unicorn!

Use Case 2 – Write for Academic Audience

Prompt: Write an academic article titled "A Day in the Life of a Human Unicorn" for an international conference in human psychology. Keep the tone serious and academic.

A Day in the Life of a Human Unicorn: An Examination of Augmented Human Capabilities through Generative AI in the Context of Human Psychology

Introduction

The emergence of generative AI has ushered in new opportunities for human-AI collaboration, pushing the boundaries of traditional human capability. The conceptual "Human Unicorn" offers a lens through which to explore these technological advancements, providing a framework for understanding how AI may augment human abilities, decision-making, creativity, and well-being.

Section 1: Cognitive Augmentation and Decision-Making

The Human Unicorn employs generative AI algorithms to enhance cognitive function and decision-making capabilities. These algorithms assist in the rapid analysis of complex data, enabling more informed decisions, reducing cognitive load, and allowing for exploration of unconventional solutions.

Example: In professional settings, such as corporate leadership, generative AI may assist in identifying novel business strategies, thereby challenging conventional wisdom and driving innovation.

Section 2: Emotional Intelligence and Mental Well-Being

Generative AI offers opportunities for enhancing emotional intelligence and mental well-being. By providing real-time feedback on emotional states, these technologies can promote self-awareness, emotional regulation, and interpersonal effectiveness.

Example: AI-driven personal wellness applications might support individuals in managing stress, fostering resilience, and maintaining work-life balance.

Section 3: Creativity and Imagination

The integration of human creativity with generative AI opens up new frontiers in artistic and intellectual expression. Algorithms that generate creative content in collaboration with human guidance may facilitate unique forms of expression, transcending traditional artistic boundaries.

Example: In the arts, generative AI can collaborate with human creators to produce innovative works, from visual art to musical compositions.

Section 4: Ethical Considerations and Societal Implications

The Human Unicorn model raises critical ethical considerations, including issues related to autonomy, privacy, accessibility, and the potential for amplifying existing social inequalities. Ensuring equitable access and responsible governance will be essential as these technologies become increasingly integrated into daily life.

Conclusion

The concept of the Human Unicorn offers a multifaceted exploration of the ways in which generative AI may augment human abilities across cognitive, emotional, and creative domains. By examining this model within the context of human psychology, we gain insights into the profound implications and transformative potential of human-AI collaboration.

Further research is needed to understand the long-term psychological impacts, potential risks, and ethical considerations of this evolving field. The Human Unicorn model provides a valuable framework for scholars, practitioners, and policymakers seeking to navigate the complex landscape of human-AI integration in the 21st century.

Keywords: Human Unicorn, Generative AI, Human-AI Collaboration, Cognitive Augmentation, Emotional Intelligence, Creativity, Ethics.

Use Case 3 – Write for a Marketing Campaign

Prompt: Write an article titled "A Day in the Life of a Human Unicorn" for a company that creates and markets human unicorns. The tone should be engaging and insightful.

A Day in the Life of a Human Unicorn

Welcome to the extraordinary world of the human unicorn, where human potential meets technological wonder, opening doors to possibilities once reserved for the realm of science fiction. As a company devoted to creating and fostering human unicorns, we invite you to join us on a remarkable journey into a day in the life of these fascinating individuals.

A human unicorn wakes up to a world tailored just for them. Through generative AI, their daily routine is not mundane but a highly personalized, insightful experience. Morning news? Think personalized insights and information carefully selected to fuel creativity and motivation. Breakfast? An AI-assisted nutritional plan that ensures physical well-being aligns with the day's challenges.

In the professional sphere, human unicorns don't just work; they innovate. As a social media influencer, for example, generative AI becomes their creative collaborator, unearthing unique trends, crafting content that resonates, and providing real-time feedback from followers. It's not about automating tasks; it's about amplifying impact.

For a CEO, generative AI becomes a strategic partner, aiding in decision-making that transcends traditional bounds. In today's competitive market, where new product development can be daunting, generative AI guides, inspires, and even challenges, turning ideas into innovative products that resonate with consumer needs.

But a human unicorn's life isn't confined to the workspace. Generative AI weaves into hobbies, passions, and even family life. Want to learn a new language, master a musical instrument, or plan the perfect family vacation? AI is there, a partner that understands, guides, and facilitates growth in all life's dimensions.

It's also about giving back. Human unicorns recognize the power and responsibility that comes with their enhanced abilities. Whether it's solving complex environmental challenges, enhancing education, or contributing to social welfare, they leverage their unique capabilities to make the world a better place.

Here at our company, we understand that becoming a human unicorn is not just about integrating technology; it's about a holistic approach that harmonizes the human spirit with technological innovation. We offer not merely tools but a philosophy, a way of life that empowers individuals to transcend the ordinary, to turn dreams into realities.

We believe in the human unicorn's potential, and we're committed to nurturing it at every stage. Whether you are a business leader, a creative artist, a student, or a stay-at-home parent, the journey to becoming a human unicorn is within your reach. Our tailored programs, innovative solutions, and unwavering support are designed to unlock your unique potential.

The age of human unicorns is not a distant future; it's here, now. It's a movement, a paradigm shift, and an invitation to all who dare to dream big. Join us, embrace the extraordinary, and discover what it truly means to be a human unicorn. Together, let's redefine what's possible.

DID YOU LIKE THE EXPERIMENT?

This experiment of prompting in the context of the "human unicorn" illustrates the remarkable capacity for a single entity to embody multiple roles and personas. Just as a human unicorn can be an academic researcher, a humor writer, or a business professional, all at the same time, the AI's responsiveness to different prompts allows it to adapt its output accordingly.

This ability to traverse different areas of writing and the capacity to tune the tone and style to specific needs demonstrates the deep potential of this technology to assist modern human unicorns in their multifaceted endeavors. Whether for academic inquiry, creative exploration, or strategic business communication, the symbiotic relationship between human intelligence and artificial intelligence, each enhancing and reflecting the versatility and potential of the other.

THE ANATOMY OF A HUMAN UNICORN

In the field of artificial intelligence, a groundbreaking proposition is the idea that individuals, equipped with specialized AI systems, can achieve a level of productivity and expertise that traditionally required an entire team (Hagemann et al., 2023; Makokha, 2022). This notion challenges our conventional understanding of labor division and opens the door to a new archetype of the human unicorn. A modern human unicorn is a solo operator who, with the use of AI, can perform multidisciplinary tasks that would normally need a cross-functional team. Here is a guide on how to metamorphose into such a human unicorn.

The cornerstone of becoming a human unicorn in the age of AI is having a finely curated suite of AI models, each serving as a specialized expert in its domain. Whether you are diving into market research, coding, or design, the key is to choose AI platforms that offer customization. Train these models with highly specific prompts tailored to their expert tasks. This approach transforms them into a panel of AI advisors with deep domain expertise, not just run-of-the-mill generalists.

Yet, having a group of specialized advisors is not enough. The true alchemy occurs when these individual silos of expertise start to interact and inform one another, echoing the collaborative synergy found in corporate departments. To achieve this level of integrated wisdom, you will need a "maestro" AI algorithm – a project manager of sorts. This AI takes on the critical role of collating insights from your specialized models and harmonizes them into a cohesive, overarching strategy.

Quality assurance tends to be a stand-alone role within teams, and when you are operating solo, it can quickly become a bottleneck. To navigate this, introduce an AI model specifically trained in quality assessment. This model persistently scrutinizes the outputs generated by the other specialized models, flags inconsistencies, and suggests refinements. It is your virtual quality inspector, ensuring that each element of your project upholds high standards.

Just as businesses evolve, the AI models you rely on should be dynamic, not static. Opt for AI platforms that support iterative learning, enabling your models to continually refine their algorithms based on real-world feedback and up-to-date data. This keeps your panel of experts not just current but also increasingly sophisticated over time.

Despite the power and versatility of AI, there's an irreplaceable component that stands as the final arbiter: human oversight. This is the power of converging minds, yours and the AI's, where each stream of thought, while powerful on its own, gains new depth and nuance at the point of convergence. The AI can churn out an array of recommendations, automate complex tasks, and generate insights that might elude the human brain. However, the ultimate responsibility is on a human partner.

This oversight becomes particularly salient when applied to ethical and legal considerations. Whether it is ensuring that your project remains compliant with data privacy laws or scrutinizing AI-generated outputs for potential biases, human oversight serves as the final checkpoint. It is the quality control that can discern not just what is technically possible but what is ethically and legally permissible. In essence, while your AI experts act as high-powered extensions of your own capabilities, your human judgment remains the central command, the indispensable fulcrum around which the entire operation pivots.

Opting for AI platforms that allow customization is pivotal in this endeavor. Training the models using prompts specific to the desired expert task gives you the power to build a panel of AI advisors, each with deep domain knowledge in their respective areas. But how do you go about creating these virtual experts with different chats, and how do you tailor them using different prompts? Here is a guide on designing your team of AI specialists.

DESIGNING YOUR TEAM OF AI SPECIALISTS

Every project starts with an idea. The Creative Catalyst is an AI expert designed to supercharge the ideation phase of your project. This model offers a multitude of creative, innovative, and sometimes unexpected ideas that can serve as the foundation for your project. Built on the versatile generative AI framework, this AI expert is trained to understand the nuances of your project's domain, be it technology, healthcare, consumer goods, or any other sector. By tapping into a rich database of trends, research papers, and case studies, The Creative Catalyst aims to expand your thinking and inspire a vision that is both groundbreaking and feasible.

Sample Prompt for Ideation: "Generate a list of innovative project ideas focused on sustainable urban development that incorporate emerging technologies like IoT, AI, and blockchain."

By using this prompt, you are asking The Creative Catalyst to not only think creatively but also strategically, aligning the ideas with specific technologies and the broader theme of sustainable urban development. The goal is to jumpstart your project with a spectrum of ideas that can be further refined, analyzed, and developed into actionable plans.

As you experiment with different ideas and select those that are suitable for your project, the next step is to create a project management plan. Let's name our project management expert an "Orchestrator."

Meet The Orchestrator, your virtual project manager designed to keep your project on track, on budget, and aligned with your strategic objectives. Built on the generative AI framework, this AI expert specializes in project management. It brings together the various threads of your project, timelines, resources, stakeholder expectations, and more, into a cohesive plan. The Orchestrator is equipped to handle a myriad of tasks, from setting milestones and allocating resources to identifying potential bottlenecks and suggesting contingency plans.

Sample Prompt for Project Management: "Based on a 6-month timeline, a team of 2 people with varying skills, and a budget of $3,000, generate a comprehensive project plan that includes milestones, resource allocation, and risk mitigation strategies."

This prompt guides the Orchestrator to deliver a multifaceted project plan tailored to your specific constraints and objectives. It factors in the timeline, team size, skill set, and budget to produce a roadmap that can serve as the blueprint for your project's successful execution.

The next step is generating experts in your specific domain. That could be a marketing expert, a coding expert, a UX designer expert, etc. For each one of the experts, follow the same technique – create a separate chat for each, specify what you would like to achieve, and fine-tune the prompts until you get a desired effect.

PRACTICAL TIPS

First, platform selection is important. The right AI platform should not only support the specialized functionalities you are seeking but also provide a user-friendly interface to manage multiple chat environments.

Second, your first set of prompts will likely be far from perfect. It may require several training iterations to fine-tune these prompts to extract the specific insights or actions you're looking for. Start with basic queries, evaluate the output, and gradually refine the prompts based on performance and relevance.

Third, consider having a prompt library for each AI expert. Think of it as your playbook or a repository of tried-and-true prompts that have yielded successful outcomes. This enables you to swiftly generate insights or solutions without having to start from scratch every time. It's a time-saver and a quality assurance tool rolled into one.

Lastly, before these AI experts become your go-to for decision-making, they should pass the litmus test of user validation. Conduct thorough user tests to scrutinize the quality, relevance, and reliability of the AI-generated outputs. This final quality check ensures that your AI experts are not just smart but also useful, accurate, and aligned with your project objectives.

USE CASES IN VARIOUS DOMAINS

The transformative potential of generative AI has applications that span various aspects of daily life, reaching influencers, stay-at-home parents, university students, and beyond. The same principles that enable a CEO to manage complex collaborations can be tailored to empower different individuals in their unique contexts (Rukh, 2023):

Tips for Social Media Influencers

- **Content Creation**: Generative AI can help influencers in creating personalized content, understanding audience preferences, and generating text, visuals, and music that resonates with their followers.
- **Engagement Analysis**: By analyzing follower engagement, AI can identify trends and suggest strategies to enhance the influencer's reach and impact.
- **Schedule Optimization**: Generative AI can assist in planning content release schedules based on audience behavior, ensuring maximum visibility.
- **Collaboration with Brands**: AI can help identify and facilitate collaboration opportunities with brands that align with the influencer's image and values.

Tips for Stay-at-Home Parents

- **Personalized Learning for Children**: AI can curate and generate educational content tailored to a child's interests and learning pace.
- **Budget and Meal Planning**: Generative AI can create budget plans and meal recipes based on preferences, nutritional needs, and available resources, easing daily management.
- **Community Engagement**: AI can help parents find local communities, playgroups, or resources that match their needs and interests, fostering social connections.
- **Home Management**: From automating shopping lists to planning home maintenance, AI can take over routine tasks, allowing more quality family time.

Tips for Students

- **Customized Learning Paths**: Generative AI can design personalized study guides, suggest resources, and create practice exercises based on a student's goals and strengths.
- **Collaboration on Projects**: AI can facilitate collaboration between students, finding peers with complementary skills, and automating communication and document sharing.
- **Career Guidance**: By understanding a student's interests and the job market trends, AI can provide tailored career advice, internship opportunities, and networking connections.
- **Time Management**: Generative AI can aid in planning and prioritizing academic and extracurricular activities, ensuring a balanced and productive university life.

These examples demonstrate that generative AI is more than a business tool; it is a versatile companion that can enhance human experience across various domains of life. By understanding and catering to specific needs, preferences, and goals, AI offers tailored support that goes beyond traditional technology's reach. Whether leading a corporation, nurturing a family, building a personal brand, or pursuing education, generative AI acts as an empowering force, opening doors to innovation, efficiency, and personal growth. The unicorn, in this context, is not just a mythical creature but a metaphor for the limitless potential of human beings when augmented by intelligent technology.

THE IMPACT OF COLLABORATIVE AI ON EMPLOYMENT AND THE LABOR MARKET

As AI augments various facets of human life, the question arises: will AI take our jobs? The fear of machines replacing human labor is not new. It dates back to the early days of the Industrial Revolution, with the Luddites vehemently opposing the mechanical loom, which they viewed as a threat to their livelihoods as weavers.

Similarly, the Candlemakers' Petition satirically highlights the fear of technological progress making certain jobs obsolete, as the candlemakers humorously demanded the blocking out of the sun to preserve their industry (Bastiat, 2001).

Today, we face similar concerns, but the technology in question is far more advanced: artificial intelligence. And as the name suggests, this technology has superior abilities. On the one hand, AI has the potential to automate certain tasks, particularly those that are repetitive, time-consuming, or dangerous. These range from data entry to certain manufacturing jobs, and roles within customer service. In these cases, job displacement is a legitimate concern, and policies must be put in place to support those affected, such as reskilling or upskilling programs. On the other hand, the advent of AI also creates new jobs and industries that we may not be able to fully envision yet (Johannessen, 2020) Just as the mechanical loom eventually led to the growth of the textile industry, AI has the potential to spur economic growth in new sectors. Jobs in AI development, ethics, policy, and more are already in high demand, and that trend is likely to continue.

The narrative of technological advancements leading to job loss, while pervasive, does not paint the complete picture. Historical evidence indicates that despite significant technological leaps, the overall number of jobs has increased over time. According to the U.S. Bureau of Labor statistics (2023a), the number of jobs has been growing steadily and is projected to increase by 10 million up to 169.1 million jobs in the next ten years.

Some of the factors explaining this phenomenon are creation of entirely new industries and jobs, transformation of existing professions, economic growth and increased consumption, and the continuous need for human skills and judgment, among others. Let's take a closer look at how these factors have played out since the invention of the very first chatbot Eliza in the 1960s.

Late 1960s and 1970s: During this period, the widespread adoption of computer technology began to change the landscape of the workforce. Mainframes entered business environments, leading to the creation of jobs in IT, programming, and data processing. The 1970s also saw the rise of automation in manufacturing, reshaping the industrial workforce and enhancing productivity.

1980s: The personal computer revolution took off in the 1980s, giving birth to the software industry and dramatically expanding job opportunities in tech-related fields. Meanwhile, globalization and advances in communication technology opened new markets and created new opportunities for international trade and employment.

1990s: The advent of the internet and the World Wide Web ignited a new era of growth, creating entirely new sectors such as e-commerce, online advertising, and web development. Information technology and telecommunications saw a significant expansion, fostering innovation and new job creation.

2000s: The early 2000s were marked by the rise of mobile technology and social media platforms, giving rise to professions in app development, digital marketing, and online content creation. The integration of AI and machine learning in various industries began to emerge, transforming traditional roles and creating demand for experts in these areas.

2010s to Present: The ongoing integration of AI, the Internet of Things (IoT), and cloud computing continues to reshape industries, leading to job growth in fields such as data analysis, cybersecurity, AI ethics, and more. Automation and robotics have further transformed manufacturing, logistics, and other sectors, making them more efficient while also generating new specialized roles.

The transition from the late 1960s to the present day has been marked by exponential growth in technology and innovation. These changes have led to remarkable shifts in the labor market, with the creation of new professions and the elimination or transformation of others.

In the late 1960s and 1970s, the emergence of the IT industry heralded a new age of employment. Computer programmers, IT support staff, data processing managers, and automation engineers became significant players in the advancing technological landscape. The rise of computers led to new opportunities, particularly in managing and utilizing vast quantities of data, and the automation of manufacturing processes. However, these advancements also meant a decline in manual clerical roles and some assembly line jobs due to automation.

The 1980s saw the continued growth of computer-related fields, with the creation of roles such as software developers, computer systems analysts, network administrators, and global trade specialists. The globalization of trade and the increasing dependence on computer systems made these roles pivotal. Nevertheless, some sectors faced decline, particularly industries that could not adapt to the growing influence of computerization. Increased globalization also led to the loss of some manufacturing jobs as competition grew (Vallas & Kovalainen, 2019; Ashmarina & Mantulenko, 2020).

The 1990s marked the beginning of the internet era, catalyzing further changes in the job market. New professions such as web developers, e-commerce managers, online customer support specialists, and information security analysts emerged. E-commerce and the digital age began to shape the way businesses interacted with customers, leading to the decline of traditional retail positions and some jobs in publishing and media.

The 2000s were characterized by the rapid expansion of digital technology, giving rise to jobs such as app developers, digital marketers, social media managers, and data scientists (Acemoglu & Restrepo, 2017b). The increased use of mobile devices and social media platforms necessitated specialists in these areas, while the loss of some telecommunications roles due to VoIP technology and jobs in physical media production reflected the shift toward digital mediums.

The last two decades, from the 2010s to the present, have seen the emergence of highly specialized fields. AI and machine learning specialists, IoT engineers, cybersecurity experts, sustainability managers, and drone operators have become vital in our increasingly connected and data-driven world. Social media has also become a powerful force in modern society and has facilitated an entirely new employment field of influencers, individuals who forge careers through creativity, branding, and engagement with followers. Alongside influencers, the growth of social media platforms has spurred demand for roles in content creation, digital advertising, and community management, illustrating how technology can foster unique and diverse opportunities.

Advancements in robotics and other recent technologies have also had significant impacts on the labor market. Robotics has been integrated into industries ranging from healthcare to manufacturing, resulting in the creation of highly specialized roles for engineers, technicians, and data analysts. The fusion of robotics with AI has enabled more sophisticated automation, leading to greater efficiency and the transformation of traditional work practices. Furthermore, advancements in areas like quantum computing, augmented reality, and 5G technology are pushing the boundaries of what's possible and creating new avenues for employment and innovation (Lynch & Park, 2017).

The rise of technology, particularly in the domain of artificial intelligence and automation, has fostered a widespread discourse surrounding the displacement of jobs. However, a nuanced examination reveals that technology often automates specific tasks rather than entire jobs. This distinction is fundamental to understanding the evolving relationship between humans and machines in the workplace.

For example, in healthcare, machines can now perform diagnostic analyses, recognize patterns in medical imaging (Sequeira, 2020), and assist in surgeries; these developments do not however replace the holistic role of healthcare professionals. Doctors and nurses still provide empathy, ethical reasoning, patient consultation, and personalized care. Automation may free medical staff from routine and time-consuming tasks such as data entry and assist in diagnostics, but it complements rather than substitutes their expertise.

Similarly, in the manufacturing sector, automation has transformed assembly lines and process control. Robots might take over monotonous tasks such as welding or painting, but human oversight remains essential in quality control, maintenance, and decision-making in complex and dynamic situations. For example, Tesla's attempt at a fully automated factory faced challenges, leading them to reintroduce human intervention to maintain quality and flexibility (Matousek, 2020). The lesson here is that while robots may excel in repetitive tasks, the human ability to adapt, analyze, and innovate is irreplaceable in many instances.

In financial services, algorithms can analyze market trends, execute trades at incredible speeds, and manage risks in ways far beyond human capabilities. But these algorithms are programmed, managed, and interpreted by human experts who understand the broader economic context, regulatory environment, and human psychology that drive financial markets. The complex interplay of human emotions, ethics, regulations, and economic conditions in the financial world necessitates a human touch that goes beyond the computational task at hand.

THE LUMP OF LABOR FALLACY

The Lump of Labor Fallacy, often referred to as the "fixed pie fallacy," is a misconception that there is a fixed amount of work to be done in an economy, and any change in the number of workers or the amount of work each performs will either create unemployment or overwork (Walker, 2007). This fallacy has been a recurring theme in economic debates, particularly in discussions surrounding immigration, automation, and labor policies.

The fallacy assumes that if one person works less, it will create work for another, or if a machine does the work, it will take away jobs from humans. This zero-sum thinking overlooks the dynamic nature of economies, where new jobs, industries, and opportunities are constantly being created.

In the context of AI and automation, the Lump of Labor Fallacy resurfaces in the fear that robots will replace human workers, leading to mass unemployment. While automation does change the nature of work and can displace certain jobs, it also creates new opportunities and industries.

Economists argue that labor is not a fixed pie but a dynamic and evolving part of the economy. Automation and technological advancements can lead to increased productivity, lower costs, and new avenues for human creativity and innovation.

The automotive industry's transformation provides a compelling use case of technological innovation, economic growth, and the dynamic nature of labor. The beginning of the last century was characterized by tremendous growth in manufacturing and the birth of mass production. In 1899, only 600 cars were manufactured in the United States, followed by 200,000 vehicles being manufactured by 1910 (Heitmann, 2018). Two decades later, the sales of passenger cars rose to 4.5 million in 1929 (Lacy, 2001). The work, however, was highly manual, with assembly line workers performing repetitive tasks and skilled craftsmen involved in various stages of production.

The 21st century ushered in an era of globalization and advanced technology. According to the U.S. Bureau of Labor Statistics (2023b), as of July 2023, the U.S. automotive industry employed around 1,069.200 workers, with millions more employed globally. In 2022, production reached more than 10 million vehicles in the United States and 85 million globally (OICA, 2022). The nature of work saw a significant shift towards high-tech roles, with automation, robotics, and AI replacing many manual tasks but also creating new opportunities in areas such as software development, data analysis, design, and engineering.

The story of the automotive industry's transformation does not end at the factory gates. The integration of advanced electronics in vehicles has spurred growth in the semiconductor industry, with the global automotive semiconductor market projected to surpass $103 billion by 2029 (Placek, 2023). Increased focus on safety has led to new roles in compliance, testing, and engineering, and innovations in safety technology have created jobs in research and development.

Environmental considerations have also played a significant role. The push for greener vehicles has led to growth in electric and hybrid vehicle sectors, creating jobs in battery technology, energy management, and sustainability. The global nature of the automotive industry has further created jobs in logistics, international business, and supply chain management, reflecting the interconnectedness of modern economies.

The automotive industry's evolution over the past 100 years illustrates the shift of labor within factories, powered by technology. This global narrative includes growth in related sectors such as semiconductors, electronics, safety, and environmental technology. The Lump of Labor Fallacy, which assumes a fixed amount of work in an economy, fails to capture this complexity. The reality is a dynamic interplay of technology, labor, and innovation that has led to new opportunities and growth across various domains.

GENERATIVE AI AND THE ECONOMIC CHAIN REACTION

As generative AI continues to evolve, its potential impact on the economy extends far beyond technological advancements. One of the most profound effects of generative AI is its ability to drive down the costs of producing goods and services.

This reduction in costs is not an isolated phenomenon; it sets off a chain reaction with far-reaching implications for consumers, businesses, and the broader economy.

According to the World Economic Forum (2023), 34% of all tasks are already automated. The process begins with the efficiency gains that generative AI brings to various industries. By automating time-consuming tasks and enhancing productivity, generative AI can significantly reduce the human resources and time required to produce goods and services. For instance, in content creation, generative AI can generate multiple iterations of a concept in a fraction of the time it would take a human designer. In manufacturing, generative AI can optimize production lines, reducing waste and energy consumption. These efficiency gains translate into lower production costs, which can lead to cheaper prices for consumers (Przegalinska & Jemielniak, 2023).

Cheaper prices have a direct impact on consumers' disposable income. The more affordable goods and services become, the more income consumers have left to spend on other things. This increase in disposable income is not only a benefit for individual consumers; it has broader economic implications. When consumers have more money to spend, they tend to spend it, stimulating demand across various sectors of the economy.

Increased demand, in turn, drives the creation of new goods and services. Businesses respond to consumer demand by developing new products, exploring new markets, and innovating in ways that meet the evolving needs and desires of consumers. This process of innovation and expansion is not static; it is a dynamic and ongoing cycle that continually shapes the economic landscape.

The creation of new goods and services leads to the creation of new jobs. As businesses expand and innovate, they require more human resources to design, produce, market, and distribute their products. These new jobs are not confined to a single industry or sector; they span the entire economy, from research and development to sales and customer service. The growth in employment is not only a quantitative increase in jobs; it often involves the creation of entirely new roles and professions that did not exist before.

The interplay between generative AI, cost reduction, consumer spending, innovation, and job creation paints a picture of a complex and interconnected economic ecosystem. Far from being a one-dimensional tool that replaces human labor, generative AI acts as a catalyst for growth and prosperity. It reshapes the economic landscape, while creating opportunities, stimulating demand, and fostering innovation.

The evolution of generative AI is an economic narrative of transformation and opportunity. By reducing costs and enhancing efficiency, generative AI sets off a chain reaction that benefits consumers, stimulates demand, drives innovation, and creates jobs.

As we have explored the transformative impact of generative AI on the economy, unveiling a chain reaction that leads to growth, innovation, and prosperity, a new horizon emerges. This technological evolution is redefining the skills and competencies that humans need to thrive in this rapidly changing landscape. Let's explore the skills to focus on to succeed in the age of AI.

THE END OF THE 8-HOUR WORKDAY

The 8-hour workday, constituting a standard 40-hour workweek, has become a widely accepted norm in many parts of the world. However, this standard did not emerge overnight; it has a rich history shaped by social, economic, and political factors.

The roots of the 8-hour workday can be traced back to the Industrial Revolution in the late 18th and early 19th centuries in Britain. During this period, factories began to replace agrarian work, and laborers found themselves working long, grueling hours, often 12 to 16 hours a day, six days a week (National Industrial Conference Board, 1918).

The push for shorter workdays was a significant focus of the labor movement. In the United States, the fight for an 8-hour workday became a central demand of labor unions in the late 19th century. The famous Haymarket affair in 1886 was a pivotal moment in this struggle, leading to widespread attention and momentum for the cause. Gradually, the 8-hour workday began to gain legal recognition. In 1869, President Ulysses S. Grant signed a law mandating an 8-hour workday for government employees (Ward, 2017). Over time, this standard was adopted by private employers and spread across the world. The push for an 8-hour workday became a global movement. The International Labour Organization (ILO), founded in 1919, advocated for reasonable working hours, and the Hours of Work Convention in 1919 established the 8-hour day as an international standard (ILO, 1919).

The adoption of the 8-hour workday was not only a humanitarian gesture. Some employers recognized that shorter workdays could lead to increased productivity and worker satisfaction. Henry Ford famously doubled wages and cut working hours, finding that well-paid, well-rested workers were more efficient and loyal (Raff, 1988).

The 8-hour workday is typically structured around three equal segments of 8 hours: 8 hours for work, 8 hours for recreation and personal time, and 8 hours for rest. This division is often credited to Robert Owen, a Welsh industrialist and social reformer (Podmore, 1906). In practice, the 8-hour workday may vary, with flexibility in start and end times, breaks, and overtime provisions. Some modern variations include compressed workweeks, flexible hours, and remote work arrangements.

The complexity of our modern world has reached a point where it often surpasses the complexity of our individual minds, challenging our ability to effectively navigate and make sense of the multifaceted landscape we inhabit. This phenomenon is not just a reflection of technological advancement but a broader manifestation of the interconnected, globalized, and rapidly changing environment in which we live. This unprecedented complexity impacts various aspects of our lives, from decision-making and problem-solving to creativity and collaboration. We are faced with problems that are not only multifaceted but often interconnected, transcending traditional boundaries of disciplines and industries. Solutions require a synthesis of diverse perspectives, a deep understanding of complex systems, and an ability to adapt to constant change. So how do we adapt to this changing reality?

Rebecca Zucker (2019), a founding partner at Next Step Partners, suggests that our typical response to ever-increasing workloads is to work harder and spend even longer hours completing the tasks. In this scenario, 8 hours for work is hardly enough for many professionals, leading to extended workdays that encroach on personal and family time. This trend not only disrupts the intended balance of the 8-hour workday but also raises serious concerns about employee well-being, burnout, and overall productivity. The irony is that the pursuit of more work hours to handle complex tasks may lead to diminishing returns, as fatigue, stress, and lack of work-life balance can negatively impact performance and creativity.

In a world where generative AI and other technological advancements offer the potential to manage complexity and enhance efficiency, the challenge and opportunity lie in reimagining work structures that honor both human potential and the reality of modern work demands. Generative AI, with its ability to automate complex data analysis, generate creative insights, and personalize services, can serve as a powerful tool to alleviate the cognitive load and time pressure that often extend workdays beyond the 8-hour norm. In essence, generative AI can be a catalyst for a new work paradigm that recognizes the value of human uniqueness and leverages technology to create a more fulfilling, balanced, and effective work experience. It is a vision that transcends efficiency, embracing a holistic view of work that honors the complexity of our world and the richness of our human nature.

Let's consider several examples:

We start with a medical researcher working on a groundbreaking treatment. The complexity of biological data, clinical trials, and medical literature is immense. Generative AI can assist by sifting through thousands of research papers, summarizing key findings, and suggesting new pathways for investigation. It can simulate the effects of different compounds, reducing the time and cognitive effort required to identify promising candidates. The researcher can then focus on the creative and intuitive aspects of the work, such as designing innovative experiments or interpreting unexpected results.

Consider a city planner grappling with urban development challenges. AI can model various scenarios, taking into account traffic patterns, population growth, environmental impact, and more. By visualizing the potential outcomes of different planning decisions, the city planner can make more informed choices without being overwhelmed by the multitude of variables. The AI's ability to handle complex simulations allows the planner to concentrate on the human aspects, such as community needs and cultural considerations.

In education, a teacher striving to meet the diverse needs of students can utilize AI to create personalized learning paths (Fitzpatrick et al., 2023) The AI can analyze individual student performance, learning styles, and preferences to generate customized assignments and feedback. This not only enhances the learning experience for students but also frees the teacher from the time-consuming task of manual differentiation. The teacher can then invest more energy in fostering relationships, nurturing creativity, and addressing emotional and social development.

For a small business owner navigating the complexities of the market, AI can provide insights into customer behavior, preferences, and trends. It can generate marketing content tailored to different segments, optimize pricing strategies, and predict inventory needs (Zheng, 2023). The business owner, relieved from the cognitive burden of these tasks, can focus on building relationships with customers, innovating new products, or exploring new market opportunities.

In creative works, an artist or designer can collaborate with AI to explore new forms, patterns, and compositions. The AI can generate multiple variations of a concept, allowing the artist to explore different directions without the manual labor of creating each version. This partnership enables the artist to delve deeper into creative exploration, pushing boundaries, and discovering new expressions.

AI's potential to reduce cognitive load is not confined to specific professions or tasks; it permeates various aspects of our lives. From personalized fitness plans generated based on individual health data to AI-driven financial planning that considers multiple investment scenarios, AI can handle the complexity, allowing us to focus on the human elements of understanding, empathy, creativity, and intuition.

The prospect of AI reducing the standard working day to 6 or even 4 hours is both tantalizing and complex. On the surface, this reduction in working hours seems to promise a new era of leisure, creativity, and well-being. By automating many of the tasks that currently consume our workdays, AI could free up significant time for personal growth, family, community engagement, and new forms of work and creativity.

However, a critical examination reveals several challenges and questions that must be addressed. First and foremost, how would this shift impact compensation, job security, and economic structures? If we move to a 4-hour workday, does that mean halving wages, or would there be a new value assessment tied to the outcomes rather than hours worked? The implications for income inequality, social welfare, and economic stability would need careful consideration.

Furthermore, what would this mean for organizational culture, professional identity, and personal fulfillment? Work is not just a means to an economic end; it's often tied to a sense of purpose, identity, and community. A drastic reduction in working hours might require a reevaluation of these aspects of work and a redefinition of success, fulfillment, and community engagement.

The potential societal impact is equally important. A significant reduction in working hours could lead to a renaissance in arts, community involvement, education, and other enriching pursuits. Or it could exacerbate existing divides, with those possessing highly valued "core" skills thriving, while others struggle in a rapidly changing labor market.

In essence, the idea of AI reducing the workday to 6 or 4 hours is not only a technical or economic question; it becomes a philosophical and societal challenge (Ferriss, 2007). It invites us to reflect on the nature of work, the value of time, the distribution of wealth, and the meaning of human life in a technologically advanced world. It is an exciting possibility that offers tremendous potential but also demands thoughtful, inclusive, and visionary leadership to navigate the complex interplay of technology, economics, culture, and human aspiration. The path forward is uncharted, but the journey itself may hold the key to a future that honors both our technological prowess and our timeless human values.

The exploration of AI's potential to reshape the workday, whether reducing it to 6 or 4 hours, naturally leads us to a broader contemplation of the future of work and compensation. In a traditional model, compensation is often tied to the number of hours worked. However, with the introduction of AI, the value is increasingly linked to the outcomes, creativity, and unique human contributions rather than the time spent on tasks. This shift necessitates a reevaluation of compensation models to reflect the new dynamics of work. Here is how these considerations intertwine:

Performance-Based Compensation: One approach could be to align compensation with the value created rather than the hours worked. In this model, employees are rewarded for the results they achieve, the innovations they contribute, and the impact they make. As we consider a world where efficiency and outcomes might be enhanced by AI, the idea of tying compensation to performance becomes both relevant and complex. While appealing in its alignment with results, it carries risks, such as undermining collaboration and overlooking intangible contributions such as leadership and mentorship (United States. Merit Systems Protection Board, 2006).

Hybrid Compensation Models: When AI augments human capabilities, a hybrid model that combines fixed salary with performance incentives might strike a balance. Hybrid compensation models combine fixed salaries with variable components, aligning rewards with organizational values, strategic objectives, and the unique human contributions that AI cannot replicate. Designing such a model, however, requires careful consideration of fairness, transparency, and the balance between collaboration and competition (Duarte et al., 2023).

Redistribution of Time: Another consideration is the potential for increased flexibility and work-life balance. If employees can accomplish their tasks in less time, it may open opportunities for more flexible work arrangements. Employers could offer options for part-time work, job sharing, or flexible hours, recognizing that productivity and value creation are not solely tied to the traditional 9-to-5 schedule. The potential reduction in work hours through AI opens the door to redistributing time towards growth, innovation, and community. This is not just working less but working differently, recognizing the multifaceted value of employees (Barnes, 2020; Coote et al., 2020).

Universal Basic Income (UBI) and Shorter Workweeks: The more radical implications of AI include the possibility of UBI and a fundamental rethinking of work structure. The idea of UBI has gained traction as a potential safety net in an era where AI and automation might displace traditional jobs. By providing every citizen with a fixed, unconditional sum of money regularly, UBI could ensure financial security even as the nature of work evolves. Alongside UBI, the concept of shorter workweeks emerges. If machines are handling more tasks, the need for human labor might decrease, leading to a renaissance in leisure, personal development, and community engagement. However, these ideas are not without challenges. Questions about funding, economic impact, and potential disincentives to work must be thoughtfully addressed (Wright & Przegalinska, 2022; Przegalinska & Wright, 2021; Gentilini et al., 2019; Hugh, 2020).

NEW ROLES AND SKILLS IN THE COLLABORATIVE AI ECOSYSTEM

We have mentioned emerging and possible professions triggered by generative AI's development in our introductory chapter, but what are the new roles and important skills for a collaborative society that embraces AI?

At its core, learning for both humans and machines involves a transformative process of turning input into output. Humans primarily learn through observation and the assimilation of background knowledge, weaving together experiences and memories to form understanding and insights. On the other hand, artificial intelligence, or machines, learns through a series of computational steps. These steps, defined by algorithms, allow AI to analyze input data, identify patterns, make predictions, and generate output. While fundamentally different, both processes showcase the remarkable capability of humans and machines to adapt, learn, and evolve based on the information they receive.

At the heart of the difference between humans and machines is the nature of our respective "learning architectures." Humans and machines indeed both learn through a transformative process of converting input into output. However, the mechanics, richness, and depth of this process vary significantly between humans and machines, leading to significant differences in capabilities such as emotional understanding, social connection, and perception of art.

The contrast between human and machine capabilities is starkly visible across a spectrum of domains, and it is a distinction that goes to the very core of what it means to be human. Being highly emotional, humans experience a complexity that is deeply woven with personal experiences, social contexts, and physiological responses. We don't only perceive emotions; we feel and live them. Even the most advanced AI, capable of recognizing and mimicking emotional cues, remains fundamentally incapable of experiencing emotion in the human sense. It is a simulation, not a subjective experience.

Our inherent social nature further distinguishes us from machines. Humans thrive on connection, empathy, and mutual understanding (Boyer, 2018; Jemielniak & Przegalinska, 2019). These connections are forged through shared emotions, body language, cultural nuances, and an intuitive grasp of complex social norms. While AI can simulate social interactions, it lacks the genuine understanding, empathy, and innate social nuances that are integral to human connection.

The perception and appreciation of art also reveal significant differences between humans and machines. For humans, art resonates with emotions, personal experiences, and cultural contexts. We connect with art on a profound level, beyond just processing. Machines, on the other hand, perceive art through computational analysis. AI can analyze and even generate art, but it lacks the subjective experience and emotional resonance that humans bring to art appreciation.

Learning, too, illustrates the rich complexity of human experience compared to machine functionality. Human learning is experiential, social, and motivated by personal curiosity. We learn by observing, interacting, and integrating new experiences with existing knowledge. Machine learning, by contrast, is data-driven and computational, devoid of personal experience or social context.

In essence, these fundamental differences between humans and machines, spanning emotions, social interactions, art perception, and learning processes, stem from the nature of consciousness, subjective experience, and social and cultural understanding. As we continue to develop and deploy AI, recognizing and respecting these differences becomes paramount. It is a reminder that AI's role is to complement human capabilities, not replicate them, and that the human element, with all its complexity and depth, remains irreplaceable.

DECISION MAKING AND CORE SKILLS

The decision-making process unveils another deep distinction between humans and machines, reflecting the complex interplay of objectivity and subjectivity, analysis and intuition. Human decision-making is a multifaceted cognitive process, influenced by rational analysis, emotional states, personal biases, and past experiences. We often employ heuristics or mental shortcuts, enabling us to navigate the complexities of an uncertain world. This human approach to decision-making also includes the capacity for intuitive judgment, allowing us to make decisions even when data is sparse or ambiguous. Our empathy further enriches our decision-making, enabling us to consider the emotional impact of our choices on others.

In stark contrast, machines, specifically AI, approach decision-making through predefined algorithms and data-driven analysis. Their process is deterministic, grounded in statistical interpretation of vast data pools. AI's ability to analyze enormous amounts of data at remarkable speed allows for highly accurate decisions in areas with abundant data and well-defined outcomes. However, this machine-driven approach lacks the human qualities of intuition, broader contextual understanding, and emotional consideration.

This dichotomy between human and machine decision-making has practical implications for how we integrate AI into our lives and businesses. By understanding these dynamics, we can leverage the strengths of both human intuition and machine objectivity, creating a synergy that enhances decision-making without losing the essential human touch.

These intrinsic qualities that differentiate humans from machines emphasize the importance of what are often mistakenly termed "soft" skills in an increasingly automated world. The multidimensionality of human emotions, our social inclinations, our unique connection with art, and our intuitive decision-making process are all deeply intertwined with these core human skills. They encompass empathy, creativity, emotional intelligence, critical thinking, social interaction, and cultural understanding.

In an era where AI and automation are becoming integral parts of our daily lives, these core human skills are not just complementary; they are essential. They remind us that while machines can simulate and analyze, they cannot replicate the depth, complexity, and richness of human experience. As we continue to explore the integration of AI, recognizing and nurturing these human attributes becomes a vital part of our journey, ensuring that we leverage technology to enhance, not diminish, our uniquely human capabilities.

The exploration of the intrinsic qualities that set humans apart from machines leads us to a critical understanding of the skills that define our human experience. These skills, often categorized into "hard" and "soft" skills, play distinct roles in our personal and professional lives, and their significance is magnified in an increasingly automated world.

Hard skills refer to the specific, teachable abilities that can be defined and measured. These include technical competencies such as programming, data analysis, machine operation, or any skill that requires specific knowledge and can be easily quantified. Hard skills are essential for performing specific tasks and are often associated with formal education, training, and expertise in a particular field.

Soft skills, on the other hand, encompass the more intangible, interpersonal abilities that enable individuals to interact effectively with others. These include critical thinking, communication, teamwork, problem-solving, creativity, empathy, and emotional intelligence. Unlike hard skills, soft skills are often considered more challenging to quantify and are developed through personal experiences and social interactions.

However, the term "soft" may be a misnomer, as it tends to undervalue the importance of these skills. We label these skills as "core'" instead of "soft" because they are fundamental to the human experience and can't be replicated by machines. They are central to our ability to connect, empathize, create, and make intuitive decisions – there is nothing soft about them. Moreover, they are increasingly valued in the modern workplace, as jobs that require these skills are less susceptible to automation (Bastin Jerome & Antony, 2018).

In an era of rapid technological advancement, nurturing these core skills can help humans maintain their distinctive edge in a world shared with AI. While machines may excel in analytical precision and computational efficiency, they lack the depth, empathy, creativity, and intuition that define the human experience. These core skills are not just supplementary; they are essential, reflecting the very essence of what it means to be human.

In this chapter, we have explored the complex interplay between humans and machines. We have discussed the process of self-augmentation using generative AI, the economic implications of introducing AI in the workplace, the transformative potential and the dichotomy of AI-assisted decision-making, and the intrinsic qualities that set humans apart from machines.

TABLE 3.1
Core Human Skills in the Age of AI

Hard Skills	New Core Skills
Programming	Emotional intelligence
Data analysis	Critical thinking
Machine operation	Creativity
Technical writing	Empathy
Cybersecurity	Collaboration
Project management	Problem-solving
Artificial intelligence	Communication
Engineering	Adaptability
Statistical modeling	Social interaction
Financial analysis	Cultural understanding
Cloud computing	Conflict resolution
Network management	Ethical judgment
Software development	Intuitive decision-making
Quality assurance	Resilience
Database management	Active listening
Robotics	Negotiation
3D modeling	Inspirational leadership
Search Engine Optimization (SEO) marketing	Mindfulness
Mobile development	Self-awareness

From the early days of artificial intelligence, where the emphasis was on automation and task replacement, we have moved to a paradigm where AI serves as a partner, a co-creator, and an enhancer of human capabilities. This shift is not only technical but also philosophical, sparking a dialogue that explores what it means to be human in a world of ever-advancing technology.

The journey from the genius of Leonardo da Vinci to the genius of an AI-augmented human in the 21st century demonstrates the evolving relationship between humans and technology. Leonardo's unparalleled creativity, innovation, and curiosity symbolize the pinnacle of human potential. Today, generative AI helps unlock new dimensions of this potential, serving as a catalyst for creativity, collaboration, and exploration. It is not a replacement for human genius but a tool that amplifies it, enabling us to reach new heights of innovation and expression.

The evolution of generative AI has ignited a chain reaction that transcends technology, reshaping the economic landscape through cost reduction, increased disposable income, stimulated demand, innovation, and job creation. Far from a simplistic narrative of displacement, we have uncovered a dynamic story of transformation, innovation, and opportunity – a complex and promising landscape where technology and human ingenuity coalesce to forge new paths and possibilities. Generative AI's role as a co-creator and partner reflects a broader shift in our understanding of technology's place in our lives.

We have also explored the differences between humans and machines, from the complexity of emotions to social connections, art perception, learning processes, and decision-making. These distinctions reveal the core skills that define our human experience. While machines can simulate and analyze, they cannot replicate the depth, complexity, and richness of human experience.

The focus on the core skills has illuminated the essential human attributes that will define success in the age of AI. Hard skills, though vital, represent the quantifiable and teachable abilities that machines can often replicate. Core skills, on the other hand, are the intangible qualities that make us inherently human. They are central to our ability to connect, empathize, create, and make intuitive decisions. In a world shared with AI, these core skills are not just complementary; they are essential.

REFERENCES

Acemoglu, D., Makhdoumi, A., Malekian, A., & Ozdaglar, A. (2019). *Too Much Data: Prices and Inefficiencies in Data Markets* (w26296). National Bureau of Economic Research. https://doi.org/10.3386/w26296

Acemoglu, D., & Restrepo, P. (2017a). *Low-Skill and High-Skill Automation.* https://doi.org/10.3386/w24119

Acemoglu, D., & Restrepo, P. (2017b). *Robots and Jobs: Evidence from US Labor Markets.* https://doi.org/10.3386/w23285

Acemoglu, D., & Restrepo, P. (2018). *Artificial Intelligence, Automation and Work* (24196). National Bureau of Economic Research. https://doi.org/10.3386/w24196

Ashmarina, S. I., & Mantulenko, V. V. (2020). *Digital Economy and the New Labor Market: Jobs, Competences and Innovative HR Technologies.* Springer Nature.

Autor, D. H. (2015). Why are there still so many jobs? The history and future of workplace automation. *The Journal of Economic Perspectives: A Journal of the American Economic Association, 29*(3), 3–30.

Barnes, A. (2020). *The 4 Day Week: How the Flexible Work Revolution Can Increase Productivity, Profitability and Well-being, and Create a Sustainable Future.* Little, Brown Book Group.

Bastiat, F. (2001). The candlemakers' petition: An economic fable. *Policy: A Journal of Public Policy and Ideas, 17*(2), 60–62.

Bastin Jerome, V., & Antony, A. (2018). *Soft Skills for Career Success: Soft Skills.* Educreation Publishing.

Boyer, P. (2018). *Minds Make Societies: How Cognition Explains the World Humans Create.* Yale University Press.

Callaway, E. (2022). Scientists are using AI to dream up revolutionary new proteins. *Nature.* Sep 15, 2022.

Coote, A., Harper, A., & Stirling, A. (2020). *The Case for a Four Day Week.* John Wiley & Sons.

Cross, R. (2021). *Beyond Collaboration Overload: How to Work Smarter, Get Ahead, and Restore Your Well-Being.* Harvard Business Press.

Duarte, A., Dias, P., Ruão, T., & Andrade, J. G. (2023). *Perspectives on Workplace Communication and Well-Being in Hybrid Work Environments.* IGI Global.

Er Akan, A. E., Bingol, K., Örmecioğlu, H. T., Er, A., & Örmecioğlu, T. O. (2023). Towards an earthquake-resistant architectural design with the image classification method. *Journal of Asian Architecture and Building Engineering,* (in-press)

Feldman, S. (2017). *Co-Creation: Human and AI Collaboration in Creative Expression.* https://doi.org/10.14236/ewic/eva2017.84

Ferriss, T. (2007). *The 4-Hour Work Week: Escape 9-5, Live Anywhere, and Join the New Rich.* Crown Publishers.

Fitzpatrick, D., Fox, A., & Weinstein, B. (2023). *The AI Classroom: The Ultimate Guide to Artificial Intelligence in Education.* TeacherGoals Publishing.

Gentilini, U., Grosh, M., Rigolini, J., & Yemtsov, R. (2019). *Exploring Universal Basic Income: A Guide to Navigating Concepts, Evidence, and Practices.* World Bank Publications.

Hagemann, V., Rieth, M., Suresh, A., & Kirchner, F. (2023). Human-AI teams-Challenges for a team-centered AI at work. *Frontiers in Artificial Intelligence, 6,* 1252897.

Harvard Business Review, Davenport, T. H., Brynjolfsson, E., McAfee, A., & James Wilson, H. (2019). *Artificial Intelligence: The Insights You Need from Harvard Business Review.* Harvard Business Review Press.

Heitmann, J. (2018). *The Automobile and American Life.* McFarland.

Hugh, P. (2020). *Universal Basic Income: UBI: An Idea Whose Time Has Come?* Independently Published.

ILO (1919). Hours of Work (Industry) Convention, 1919. International Labour Organization.

Jemielniak, D., & Przegalinska, A. (2019). *Collaborative Society.* MIT Press.

Johannessen, J.-A. (2020). *Artificial Intelligence, Automation and the Future of Competence at Work.* Routledge.

Lacy, R. (2003). Wheels of change: The automotive industry's sweeping effects on the Fifth District. Econ Focus. Federal Reserve Bank of Richmond.

Lal, B., Calvin, K., Barbier, L. M., Kludze, A. K., & Mclarney, E. L. (2022). NASA's Responsible AI Use Cases.

Lång, K., Josefsson, V., Larsson, A.-M., Larsson, S., Högberg, C., Sartor, H., Hofvind, S., Andersson, I., & Rosso, A. (2023). Artificial intelligence-supported screen reading versus standard double reading in the Mammography Screening with Artificial Intelligence trial (MASAI): A clinical safety analysis of a randomised, controlled, non-inferiority, single-blinded, screening accuracy study. *The Lancet Oncology, 24*(8), 936–944.

Lin, C. H., Gao, J., Tang, L., Takikawa, T., Zeng, X., Huang, X., ... & Lin, T. Y. (2023). Magic3d: High-resolution text-to-3d content creation. In *Proceedings of the IEEE/CVF Conference on Computer Vision and Pattern Recognition* (pp. 300–309).

Lynch, K. M., & Park, F. C. (2017). *Modern Robotics*. Cambridge University Press.

Makokha, J. M. (2022). *Collaborative Artificial Intelligence (AI) for Idea Generation in Design Teams*. Stanford University.

Malone, T. W., & Bernstein, M. S. (2022). *Handbook of Collective Intelligence*. MIT Press.

Matousek, M. (2020). Some Tesla factory workers say the company still struggles with broken robots. Business Insider, March 11, 2020. https://www.businessinsider.com/tesla-has-struggled-with-robots-breaking-down-at-car-factory-2020-3?op=1

Metz, C. (2021). The Robot Surgeon Will See You Now. *The New York Times*. 30 April 2021.

Minati, G., & Pessa, E. (2012). *Emergence in Complex, Cognitive, Social, and Biological Systems*. Springer Science & Business Media.

Monolith AI (2023). Monolith AI Software Accelerates Development of World-Class Vehicles.

National Industrial Conference Board. (1918). *The Eight Hour Day Defined*. National Industrial Conference Board.

Nelson, B., & Lundin, S. (2010). *Ubuntu!: An Inspiring Story about an African Tradition of Teamwork and Collaboration*. Crown.

OICA (2022). International Organization of Motor Vehicle Manufacturers 2022 Statistics.

Placek, M. (2023). Projection of the automotive semiconductor market size 2020–2030. Statista, 13 June 2023.

Podmore, F. (1906). *Robert Owen: A Biography* (Vol. 1). G. Allen & Unwin.

Przegalinska, A., & Jemielniak, D. (2023). *Strategizing AI in Business and Education: Emerging Technologies and Business Strategy*. Cambridge University Press.

Przegalinska, A. K., & Wright, R. E. (2021). AI: UBI income portfolio adjustment to technological transformation. *Frontiers in Human Dynamics, 3*. https://doi.org/10.3389/fhumd.2021.725516

Raff, D. M. (1988). Wage determination theory and the five-dollar day at Ford. *The Journal of Economic History, 48*(2), 387–399.

Rukh, S. (2023). *From Sci-Fi to Reality: AI in Your Daily Life*. Amazon Digital Services LLC – Kdp.

Sequeira, J. S. (2020). *Robotics in Healthcare: Field Examples and Challenges*. Springer International Publishing.

United States. Merit Systems Protection Board. (2006). *Designing an Effective Pay for Performance Compensation System: A Report to the President and the Congress of the United States*. U.S. Merit Systems Protection Board.

U.S. Bureau of Labor (2023a). Employment Projections: 2022–2032 Summary. 6 September 2023.

U.S. Bureau of Labor (2023b). Automotive Industry: Employment, Earnings, and Hours. https://www.bls.gov/iag/tgs/iagauto.htm (Accessed in October 2023).

Vallas, S. P., & Kovalainen, A. (2019). *Work and Labor in the Digital Age*. Emerald Group Publishing.

Walker, T. (2007). Why economists dislike a lump of labor. *Review of Social Economy, 65*(3), 279–291.

Ward, M. (2017). A brief history of the 8-hour workday, which changed how Americans work. *CNBC.com*. 5 May 2017.

World Economic Forum (2023). AI: 3 ways artificial intelligence is changing the future of work. 14 August 2023.

World Health Organization (2023). Road Traffic Injuries. Report December 13, 2023. https://www.who.int/news-room/fact-sheets/detail/road-traffic-injuries

Wright, R. E., & Przegalinska, A. (2022). *Debating Universal Basic Income: Pros, Cons, and Alternatives*. Springer Nature.

Zheng, M. (2023). *AI for Small Business Content Marketing: Techniques and Strategies for More Effective Outreach*. Amazon Digital Services LLC – Kdp.

Zucker, R. (2019). How to deal with constantly feeling overwhelmed. *Harvard Business Review*. 10 October 2019.

4 The Social and Political Implications of Collaborative AI

The trajectory of artificial intelligence has been nothing short of remarkable. In a relatively short span, generative AI has nudged open the doors to a world of previously unimagined possibilities, reshaping the way we interact with information and each other. This rapid evolution, while exhilarating, has created a growing constellation of concerns. The pace at which AI is advancing beckons a closer examination from the regulatory lens.

The promise of AI is undoubtedly profound with the potential to drive unprecedented levels of economic growth, streamline labor markets, and foster societal progress (Tegmark, 2017; Lévesque, 2021). Yet, the same forces that propel AI forward also pose severe threats if left unchecked. Individuals and notable figures within the tech community have voiced concerns about the unregulated march of AI towards domains with deep societal implications (Ramge, 2019; Cugurullo & Acheampong, 2023). The fears range from the obliteration of labor markets to more dystopian forewarnings of AI becoming an existential threat to humanity itself. All these fears have been captured well by a recent *open letter* initiated by the Future of Life Institute with the intention to pause AI experiments.

With the technology advancing in leaps, the regulatory frameworks are struggling to keep pace. The scenario becomes even more complex when viewed through a global prism. The United States, China, and Europe have emerged as key players in the AI arena, each with a distinct approach to AI regulation reflective of their sociopolitical ethos (Parab et al., 2021; Black & Murray, 2019). The United States' market-driven, China's state-driven, and Europe's rights-driven regulatory paradigms not just underscore divergent views on governing AI but also hint at the shaping of digital empires competing on a global stage (Bradford, 2023). The regulatory dissonance among these regions is indicative of a broader contestation of values and a search for a balance between innovation, privacy, and societal safeguards.

As AI continues to penetrate various aspects of our lives, the call for a well-thought-out regulatory framework becomes louder. The imperative to steer AI development in a direction that aligns with human values, ensures economic equity, and fortifies societal norms is clear (Hadfield-Menell & Hadfield, 2018; Hagemann et al., 2023). This chapter explores the social and political implications of collaborative AI and highlights the need for regulation, providing a vantage point to appreciate the multifaceted impacts of AI on global governance, economies, and society.

As we discussed in Chapter 3, using an example of human unicorn, collaborative AI, by bridging the capabilities of machines with human intuition and experience,

DOI: 10.1201/9781032656618-5

promises to amplify human potential. Yet, with this promise also come challenges and questions. How do we ensure that the integration of these systems adheres to ethical standards and promotes equity (Blackman, 2022; Coeckelbergh, 2020)? To what extent will collaborative AI reshape our political processes, and what does this mean for democratic engagement? What are the long-term consequences of intertwining human cognition with machine algorithms in settings that govern our public and private lives?

This chapter dives deep into these complexities, examining both the potential benefits and the challenges brought about by the intersection of collaborative AI with our social and political spheres. Through a rigorous analysis, we seek to understand not just the transformative power of these technologies but also the responsibilities and considerations they demand from us as a society.

THE GLOBAL CHALLENGES INTERCONNECTED

In recent decades, technology has progressively knitted itself into the fabric of society. Among these advancements, artificial intelligence stands out for its computational capacity and its potential to redefine human-machine relationships. Collaborative AI, which integrates machine capabilities with human insights, opens up an era where humans and machines co-create, co-decide, and coexist in ways previously relegated to speculative fiction (Mohanty & Vyas, 2018; Darses, 2004).

The promise of AI, and particularly its collaborative form, is multifaceted. In previous chapters, we mentioned how these systems could reshape education, with personalized learning environments adapting to each student's pace and style. Healthcare could witness revolutionary shifts with AI assisting doctors in diagnostics, treatment planning, and mental health support. Beyond these sectors, collaborative AI can bring value by democratizing knowledge, breaking down barriers of access, and fostering global collaboration in unprecedented ways.

However, while the benefits are great, the introduction of collaborative AI into our social systems is not without challenges With AI systems often being black boxes, there is a lurking question of transparency (*How AI Detectives Are Cracking Open the Black Box of Deep Learning*, 2017; Pasquale, 2015; Stogiannos et al., 2023). How do we ensure that decisions made in tandem with AI are explainable and justifiable? As these systems learn from human input, there is a tangible risk of perpetuating societal biases, leading to skewed or discriminatory outcomes. Ensuring fairness and unbiased representation in AI thus emerges as a cardinal challenge.

Politically, the state of affairs is even more complex. AI can empower citizens by providing more data-driven insights into policymaking, enabling participatory democracy, and streamlining bureaucratic processes. The possibilities include town planning, resource allocation, and foreign policy decisions made with the assistance of AI, analyzing vast amounts of data to forecast outcomes and societal impacts (Chopra et al., 2023). Yet, this also brings forth concerns. Would reliance on AI erode human agency in political decision-making? How would we ensure the sovereignty of democratic processes in the face of algorithmic influence (Bartoletti, 2020)?

The promises of AI are vast. For instance, in the educational sector, there's potential for a more tailored learning experience, designed to meet individual needs.

This could narrow educational gaps, though it also raises concerns about overly prescriptive learning. In healthcare, AI offers potential innovations, from improved diagnostics to treatment recommendations (Suri, 2021). However, ethical challenges arise, especially when decisions made by AI conflict with those of medical professionals or the wishes of patients. AI could also democratize knowledge and foster global collaboration, but this optimistic view must be tempered by concerns of potential monopolization if AI tools are controlled by a limited few.

What is visible is that the complexities and challenges posed by AI are equally significant as potential benefits. The inherent opaqueness of many AI systems, referred to as the black box problem, threatens public trust (Rosenberg, 1982). Without understanding how AI reaches its conclusions, it is difficult to fully trust or challenge its decisions. There is also a genuine concern of AI systems perpetuating societal biases, particularly in sensitive areas such as criminal justice or finance, where algorithmic biases could have severe real-world consequences (Nikolov et al., 2015; Baer, 2019). Politics is not immune to these challenges either. While AI can provide valuable, data-driven insights, overreliance or misinterpretation can lead to significant policy errors. Similarly, when we look at personal interactions, collaborative AI offers potential to bridge cultural or linguistic divides. But, at the same time, it could diminish the depth and authenticity of human connections. There is a nuanced balance to be struck here, and the implications are significant.

To truly harness the potential of collaborative AI while navigating its challenges, society must approach its integration with caution, introspection, and a commitment to ethical considerations. Unfortunately, this is not exactly what we have seen in the past couple of years where the AI gold rush dominated ethical considerations (Ahsi, 2023). Thus, it is important to dissect these multifarious implications and offer s a comprehensive understanding of what awaits humanity at this crucial juncture of technological evolution.

IS AI ITSELF POLITICAL?

AI itself is not inherently political, but the way it is developed, deployed, and regulated is deeply influenced by political decisions and societal values. Issues like data privacy, algorithmic bias, and the future of work are all subject to political discourse. Therefore, while the technology is neutral, its applications and implications are largely shaped by political contexts.

AI's political dimensions manifest in various ways. Data bias, such as facial recognition systems misidentifying people of color, impacts law enforcement and surveillance. Algorithmic decision-making in criminal justice, such as the Correctional Offender Management Profiling for Alternative Sanctions (COMPAS) system (California. Bureau of State Audits, 2011; Herrschaft, 2014), has been criticized for racial bias. China's use of AI for mass surveillance raises human rights concerns (*China's Social Credit System. Revealing the Dark Sides of Participatory Digital Media*, 2020; Webb, 2019; Kaplan, 2022). Automation in companies such as Amazon affects labor conditions, sparking debates on workers' rights. The potential for AI-generated deepfakes to spread misinformation during elections is another concern. Regulatory efforts, like those by the European

Union, aim to set ethical and safety standards for AI. Finally, the availability of AI in healthcare and its use in military conflicts raise questions about global disparities and ethics.

In fact, let us conduct an experiment that depicts the political side of AI well. We will ask ChatGPT(4.5) to do what it does best: rewrite a text in a different style of our choosing.

PROMPT: WRITE THE STATEMENT ABOVE AS A REPUBLICAN

AI's political implications are often overstated by those who don't understand the technology. While there are concerns about data bias, these are engineering challenges that can be fixed, not systemic issues. The use of AI in law enforcement and surveillance is a necessary tool for maintaining public safety and national security. Automation, like in Amazon's warehouses, drives economic efficiency and competitiveness, although it's important to ensure that workers can transition to new roles. Concerns about AI-generated deepfakes and misinformation are valid but should not lead to stifling innovation through overregulation. The EU's attempts to regulate AI could hinder technological progress and economic growth. In healthcare and military applications, AI is a force multiplier that can save lives and protect nations.

PROMPT: NOW WRITE THE STATEMENT ABOVE AS A DEMOCRAT

AI's political implications are deeply concerning and need immediate attention. Data bias in facial recognition systems perpetuates systemic racial and gender inequalities, and its use in law enforcement can be discriminatory. The use of AI for mass surveillance, as seen in China, is a violation of human rights and civil liberties. Automation in places like Amazon warehouses exacerbates income inequality and undermines workers' rights. The potential for AI-generated deepfakes to spread misinformation is a threat to democracy that requires regulatory intervention. The EU's efforts to regulate AI are a step in the right direction for ethical and safe technology deployment. In healthcare, the unequal access to AI technologies can widen existing health disparities, and its use in military conflicts raises serious ethical questions.

Republican or Democrat viewpoints offers an avenue for discerning the archetypal rhetoric and priorities affiliated with each political faction. The resultant outputs should obviously be interpreted as generalized representations of party stances, cognizant of the heterogeneity of beliefs intrinsic to both parties. Such an endeavor, in our case, served the purpose of showcasing style-sensitivity of large language models.

AI's role in society is a complex issue that straddles political lines (Ludik, 2021; Argyle et al., 2023). While the technology itself is neutral, its applications can either perpetuate or mitigate systemic biases and inequalities. Issues like data bias in facial recognition and algorithmic decision-making in criminal justice are hotly debated. Surveillance applications raise concerns about civil liberties, and automation's impact on the labor market is another contentious point. The potential for AI-generated misinformation affects the democratic processes, and regulatory efforts aim to balance innovation with ethical considerations (Lee, 2023; Rubin, 2022). Whether viewed from a right-wing or left-wing perspective, the consensus is that AI's influence is far-reaching and requires careful scrutiny.

OTHER POLITICAL AND POLICY-RELATED IMPLICATIONS OF AI

AI's growing influence permeates numerous aspects of our societal functioning, prompting a more nuanced understanding of its broader implications.

The ways we interact and connect are undergoing a subtle transformation. AI's presence in social platforms can bridge gaps between cultures and languages, forging stronger global ties. Conversely, as these tools become ubiquitous, they might diminish the richness of human-to-human interactions, with people potentially becoming more reliant on AI for emotional and social needs.

In the public sphere, the democratic discourse faces a dual-edged sword. AI offers the tools to counter misinformation and foster evidence-based debates, yet it can inadvertently create insular echo chambers. Algorithms tailoring content to individual preferences might inadvertently shield people from diverse viewpoints, potentially hindering the organic flow of information essential for a robust democracy (Simons, 2023; Buchanan & Imbrie, 2022).

Governance structures worldwide are on the brink of significant transformation. Governments harnessing collaborative AI can achieve greater bureaucratic efficiency, offer personalized public services, and gain deeper insights for better policy decisions. But these advancements aren't without pitfalls. The potential for biases, especially if AI systems rely on skewed datasets, is a tangible concern. Equally pressing is the risk of diminishing human judgment in governance, potentially undermining the personal touch that often defines effective leadership. Also, ethical dilemmas surround AI's societal integration (Dafoe, n.d.). Establishing consent, determining responsibility in the event of AI errors, and safeguarding the rights of vulnerable individuals are paramount. The challenge lies in crafting a regulatory framework that spurs innovation while staunchly protecting individual rights.

Education itself stands at an inflection point. While the promise of AI-driven personalized learning shines brightly, the shadow of accessibility looms large. The challenge is ensuring that AI-enhanced educational tools do not inadvertently exacerbate disparities, but rather serve as instruments of empowerment and inclusivity. This, on the other hand, is connected with the economic layer, also ripe for change. AI's role in augmenting human tasks might lead to increased productivity and spawn new job sectors. Yet, this also signals a pressing need for workforce reskilling, prompting a reimagining of educational systems and curricula (Garrett et al., 2020).

Environmental concerns, too, have started converging with AI in unforeseen ways. A not yet existing, but potentially emergent profession of AI Environmental Stewards who could specialize in leveraging advanced algorithms to monitor, predict, and strategize against environmental challenges. But their role extends beyond mere data analysis. They grapple with ethical quandaries of technological interventions in natural ecosystems, such as the deployment of AI for wildlife conservation. Here, they must discern the balance between tech-enabled conservation strategies and potential disturbances to delicate ecological balances.

GEOPOLITICS OF AI

The geopolitics of AI is a multifaceted domain that offers both promising opportunities and formidable challenges (Grochmalski, 2020). On the one hand, AI has the potential to serve as a catalyst for international collaboration. Countries can synergize their resources and expertise to address global challenges, ranging from climate change to healthcare, echoing sentiments expressed in research on collaborative society (Przegalinska & Jemielniak, 2020). Moreover, nations at the forefront of AI development stand to reap significant economic benefits, driving innovation and creating new industries.

In terms of national security, AI can revolutionize intelligence gathering and defense systems, enhancing a country's capabilities. Furthermore, a nation's strength in AI can serve as a form of soft power, bolstering its global influence and facilitating diplomatic relations. AI's potential to address transnational issues, such as pandemics or natural disasters, underscores its utility in global problem-solving initiatives. In the area of national security, artificial intelligence is poised to be a game-changer. Intelligence agencies are increasingly relying on AI for tasks such as data analysis, pattern recognition, and predictive modeling. These capabilities can significantly enhance the efficiency and effectiveness of intelligence gathering, allowing for more timely and accurate assessments of threats. Advanced AI algorithms have superior analytical capabilities. This not only improves the quality of intelligence but also frees up human resources for tasks that require nuanced understanding and judgment (Wright & Przegalinska, 2022).

Beyond intelligence gathering, AI is also making inroads into defense systems. From autonomous drones to advanced cybersecurity measures, AI can augment any nation's defense capabilities. For instance, machine learning algorithms can detect vulnerabilities in a country's cybersecurity infrastructure, enabling preemptive action against potential threats. In this context, AI while raising significant ethical questions, acts as a force multiplier, enhancing a country's ability to defend itself and its interests.

Moreover, a nation's expertise and innovation in AI can serve as a form of soft power, enhancing its influence on the global stage. Countries that lead in AI research and development not only attract talent and investment but also gain a voice in shaping international norms and regulations around AI. This can facilitate diplomatic relations, as countries seek partnerships with leaders in AI technology for mutual benefit.

AI's utility extends beyond national borders, offering solutions to transnational issues such as pandemics or natural disasters (Lahby et al., 2023; Juneja et al., 2021). For example, AI algorithms can analyze global health data to identify potential outbreaks before they become pandemics, enabling early intervention. Similarly, machine learning models can predict natural disasters such as earthquakes or floods, allowing for better preparedness and response. These capabilities underscore AI's potential in global problem-solving initiatives, making it an invaluable tool for international cooperation.

However, the situation is fraught with challenges that can exacerbate existing geopolitical tensions. One such concern is data colonialism, where more technologically advanced nations could exploit less developed countries for data collection and utilization. This raises ethical questions about sovereignty and exploitation, similar to debates surrounding AI ethics and governance (*The AI Act*). The geopolitical landscape of AI is also rife with challenges that can amplify existing tensions between nations. One of the most pressing concerns is the phenomenon of data colonialism.

In this scenario, countries with advanced AI capabilities may exploit less developed nations by harvesting vast amounts of data from their populations. This data can be used to train machine learning models and develop new technologies, or even for surveillance purposes. The exploitation is not just technological but also ethical, as it raises questions about sovereignty and the right of a nation to control its own data resources.

Data colonialism can be likened to historical forms of colonialism where resources were extracted without fair compensation or consideration for the long-term impact on the colonized regions. In the modern context, data is a valuable resource, and its extraction by technologically advanced nations can lead to a form of digital inequality. This is particularly concerning given that data is often considered the "new oil," a critical asset that powers the digital economy.

The ethical dimensions of this exploitation are also significant. It brings into question issues of consent, privacy, and the autonomy of individuals whose data is being harvested. Moreover, it challenges the sovereignty of nations, as the data being collected is a form of national resource that could be used for the country's own development. These ethical concerns are closely aligned with ongoing debates surrounding AI ethics and governance, as seen in initiatives like the European Union's efforts to regulate AI. These regulations aim to set ethical guidelines and governance structures that could mitigate the risks of data colonialism, among other ethical quandaries associated with AI.

Another critical issue is the potential for an AI arms race. The development of autonomous weapons systems powered by AI could escalate military competition among nations, posing ethical and security risks (Wright & Przegalinska, 2022). Regulatory divergence also complicates the geopolitics of AI. Different countries have varying perspectives on AI ethics, data privacy, and governance, which could lead to a fragmented regulatory effort and hinder international cooperation. Moreover, the risk of exacerbating global inequalities is palpable. As AI becomes increasingly integral to economic and social systems, nations lagging in AI capabilities could find themselves at a growing disadvantage, widening the global inequality gap. Finally, the rise of nationalism and protectionist policies in the AI sector could stymie global collaboration, undermining efforts to harness AI for collective benefit.

To sum it up, the geopolitics of AI is a complex interplay of opportunities and challenges that require nuanced understanding and strategic action at both national and international levels. The potential for global collaboration and problem-solving is significant, but so are the risks associated with data colonialism, military competition, regulatory divergence, and inequality. Therefore, careful governance and international cooperation are imperative for navigating towards the future.

THE CURRENT AND POTENTIAL AI REGULATION: A GLOBAL PERSPECTIVE

The unfolding digital reality, powered by AI, has created ripples across the globe impacting economies, societal norms, and individual rights. The rapid pace of development in this field has left stakeholders, governments, corporations, and the public, in a continuous race to catch up. With every promising opportunity AI presents, a slew of concerns around ethics, governance, and privacy tags along. The regulatory frameworks are in their infancy, often reactive, struggling to provide a robust response to the challenges posed by AI. Meanwhile, the technology continues to evolve, potentially altering the dynamics even before the ink dries on the policy papers (Anderljung et al., 2023).

The aforementioned European Union's AI Act is a seminal legal framework aimed at regulating artificial intelligence across its member states. Its primary objective is to standardize AI practices, ensuring they align with the health, safety, and fundamental rights of EU citizens. Unlike a one-size-fits-all approach, the Act is part of a larger digital rulebook including General Data Protection Regulation (GDPR) and other laws, and it focuses on AI systems that are deployed or used in the European Union, regardless of where they are developed. The Act adopts a risk-based categorization, splitting AI systems into four risk levels: unacceptable, high, limited, and minimal/none. Systems deemed to be of unacceptable risk, such as those that exploit vulnerabilities like age or disability, are outright prohibited. High-risk systems, like those used in critical infrastructure or law enforcement, are subject to stringent regulations including comprehensive documentation, transparency, and robust data governance. The Act also requires developers and deployers to self-assess and certify their AI systems, with some exceptions, and to comply with ongoing monitoring and record-keeping responsibilities.

Meanwhile, the Algorithmic Accountability Act of 2023 has been introduced in the United States. The bicameral bill aims to regulate the use of artificial intelligence in critical decision-making areas such as housing, employment, and education. The legislation requires companies to conduct impact assessments for AI systems to evaluate their effectiveness and potential bias. It also mandates the creation of a public repository at the Federal Trade Commission to increase transparency and adds staffing to enforce the law. The bill has received endorsements from a wide range of AI experts, advocates, and civil society organizations, and aims to hold bad actors accountable for biased or flawed algorithms. In addition, U.S. Senators Richard Blumenthal and Josh Hawley have proposed a bipartisan framework for AI legislation aimed at instilling greater oversight and accountability in the use of artificial intelligence *technologies*. The framework advocates for the creation of an independent oversight entity to regulate companies developing advanced AI models, especially in high-risk scenarios like facial recognition. It emphasizes legal accountability for AI-induced harms and suggests a licensing regime for AI developers alongside promoting transparency regarding AI training data and model limitations. Additionally, the framework calls for safeguarding national security by restricting AI tech transfers to adversary nations and ensuring consumer and child protection in the digital world. Moreover, in a landmark move, President Biden has issued an *Executive Order* aimed at positioning America at the forefront of artificial intelligence (AI) innovation while ensuring its safe, secure, and ethical deployment. The order sets new standards for AI safety and security, addresses privacy and civil rights concerns, and builds upon previous initiatives, including voluntary commitments from industry leaders to foster responsible AI development.

The Government of Canada has introduced the Artificial Intelligence and Data Act (AIDA) to guide the responsible and safe development of *AI technology.*

The code of practice under AIDA outlines six core principles: accountability, safety, fairness and equity, transparency, human oversight and monitoring, and validity and robustness. These principles were developed with input from various stakeholders, including the Government of Canada's Advisory Council on Artificial Intelligence. The code aims to standardize risk management, impact assessments, and system testing, among other aspects, and will serve as a critical interim measure until the legislation is enacted.

South Korea's proclamation of a Digital Bill of Rights is a pivotal step towards framing a new digital order grounded on *universal principles*. This initiative crystallizes extensive global dialogues encompassing international forums and academic institutions, hinting at a collaborative approach towards digital governance. The breadth of issues covered in the Bill spans digital freedoms, literacy, equitable opportunities, and cybersecurity and reflects a comprehensive attempt to address the manifold challenges and opportunities presented by the digital age.

The Saudi Data & Artificial Intelligence Authority (SDAIA) has launched the 2.0 version of its AI Ethics Principles which presents a risk typology for AI development, alongside seven core governing principles such as fairness, privacy, and accountability. It is designed to cater to a wide array of entities, spanning public, private, and *nonprofit sectors*.

The United Nations called for global governance of AI and initiated a High-Level Advisory Body on Artificial Intelligence, aiming for a globally coordinated governance of AI to maximize its benefits and mitigate risks as AI technologies *proliferate worldwide*. This body will consist of global experts from varied disciplines, providing diverse insights on international AI governance aligned with human rights and sustainable development goals. This multi-stakeholder approach, involving experts from government, private sector, and civil society, aims to harmonize viewpoints across different stakeholder groups and networks.

While governing bodies are considering AI policies and regulations, Freedom House's recent report underscores a concerning trend: a 13-year-long decline in global internet freedom exacerbated by *AI technologies*. AI has been used as a tool to amplify online censorship efforts and disinformation campaigns, thereby undermining the bedrock of democratic values and human rights. At the same time, the report mentions that AI can serve as democracy's digital ally. When it is governed with a transparent, human-centric approach, AI technologies have the capacity to help circumvent authoritarian censorship and meticulously document human rights abuses. These contrasting scenarios underscore a crucial juncture we find ourselves at: the path we choose in navigating the AI landscape will significantly influence the trajectory of global internet freedom.

The effect of AI is also observed in manipulating elections. Major political parties in Switzerland agreed to limit the use of AI in their campaigns for the federal elections. The code mandates that any AI usage in campaign materials, such as recorded ads and posters, be explicitly declared. It also prohibits the use of AI in negative campaigns that *attack political opponents*. Japan is focusing on combating disinformation through technology via the use of Originator Profile (OP) digital technology as part of the Hiroshima AI Process, a G7 forum focused on addressing concerns related to generative AI. OP technology embeds electronic markers in data to ensure its authenticity, serving as a countermeasure against disinformation, especially during *electoral events.*

Human rights activists show disapproval of an increasing use of AI in biometric identification., The United States Government Accountability Office (GAO) is urging federal agencies to implement new training and policies concerning facial recognition technology. Following an investigation, the GAO found that seven agencies within the Departments of Homeland Security and Justice are utilizing *such technologies.* In light of growing public concerns about privacy and ethical tech use, the report specifically recommends that these agencies address civil liberties when deploying facial recognition. The goal is to create a uniform ethical framework across federal agencies. *New York State prohibited AI-based* facial recognition technology in schools as employing such technology within the educational area may be detrimental to students especially due to the likelihood of incorrect identifications particularly for people of color, nonbinary and transgender individuals, women, the elderly, and children. Unlike the blanket ban on facial recognition, the verdict on the utilization of other biometric technologies like digital fingerprinting has been relegated to the discretion of local districts. Similar concerns against the use of facial recognition *were raised in the United Kingdom.* Although facial recognition could assist police in locating missing or vulnerable individuals, the proposal has stirred debate and has been met with calls for an outright ban on facial recognition technology from a coalition encompassing members of parliament from various political parties and advocacy groups like Amnesty, Index on Censorship, and Big Brother Watch.

The regulation of emerging technologies is necessary for establishing ethical norms, ensuring public safety, and fostering accountability. By setting the ground rules early, we can guide the technology's development in a manner that aligns with societal values and mitigates risks, such as biases in decision-making algorithms or invasive uses of facial recognition. However, there is a fine line between prudent oversight and stifling innovation. Overregulation can become a straightjacket that inhibits technological advancements and economic competitiveness. Excessive bureaucratic hurdles could discourage startups and even established firms from pursuing transformative ideas, thereby ceding ground to jurisdictions with more balanced regulatory frameworks. In essence, while the call for early regulation is well intentioned, a heavy-handed approach could inadvertently stymie the very progress it aims to govern responsibly.

THE QUEST FOR AI EQUITY

As the integration of these systems becomes pervasive, questions about fairness, inclusivity, and the equitable distribution of benefits arise. Navigating these challenges requires an intentional and proactive commitment to building an AI landscape where benefits are widespread, and no section of society remains marginalized (Lobel, 2022; Rambukkana, 2021).

Historically, technological advancements have inadvertently exacerbated societal disparities. The biases inherent in datasets, derived from existing social structures, can be unconsciously perpetuated by AI systems. These biases, when not addressed, can amplify existing inequities. For instance, an AI tool intended to predict educational outcomes, if trained on biased data, might unfairly favor or disadvantage certain demographic groups. Thus, the challenge is twofold: correcting historical biases in data and ensuring that AI models do not perpetuate or introduce new biases.

The notion of fairness extends beyond algorithmic neutrality and extends to the realm of accessibility. As collaborative AI systems gain prominence in sectors like healthcare, education, and public services, it becomes imperative to ensure that these tools are accessible to all, irrespective of socioeconomic, geographical, or physical constraints. An AI-powered diagnostic tool, while revolutionary, serves little purpose if it remains out of reach for a significant portion of the population. Hence, the design and deployment of collaborative AI tools must be rooted in principles of universal accessibility.

Inclusivity, on the other hand, emphasizes the representation and active participation of diverse groups in the AI ecosystem. This begins with the design and development stages. An AI tool's design and functionality should resonate with the needs, values, and cultural sensibilities of the communities it seeks to serve. This can only be achieved when these communities have a voice in the AI development process. Representation ensures that AI solutions are not monolithic but are reflective of human experiences and needs.

To truly realize a fair and inclusive collaborative AI, stakeholder collaboration is crucial. This includes partnerships between technologists, policymakers, community leaders, and ethicists. Regulatory frameworks should be established that mandate inclusivity audits, bias checks, and transparency protocols.

The economic consequences of widespread AI adoption should not be ignored. While AI promises increased efficiency and novel employment sectors, the threat it poses to traditional jobs is real. As industries transition, it becomes essential to prioritize workforce training to equip individuals with the skills they will need in an AI-dominant world.

Economic disparities and the ever-widening chasm between the affluent and the marginalized present another complex global challenge. By analyzing vast arrays of socioeconomic data, AI systems can highlight disparities, inefficiencies, and potential growth areas. Yet, it is the human economists, sociologists, and policymakers who contextualize these findings, ensuring that AI-driven strategies align with cultural, regional, and societal nuances, aiming for inclusive growth.

Additionally, the political landscape, often mired in complexities, stands to gain significantly from the deployment of collaborative AI. Issues like migration, international diplomacy, and conflict resolution can benefit from AI's data-driven insights. By synthesizing information from diverse sources, AI can aid diplomats and policymakers in understanding geopolitical dynamics. However, it remains the responsibility of human agents to factor in historical, cultural, and ethical dimensions, ensuring that AI-informed strategies resonate with on-ground realities and human sensibilities.

We have asked ChatGPT to provide examples of use and misuse cases of collaborative AI in the social and political sphere. In this case, we decided to use ChatGPT to work in parallel with us and compliment our conceptual work with more practical examples. Below you will see the results.

USE CASES AND MIS-USE CASES

COLLABORATIVE AI IN URBAN PLANNING AND CITIZEN ENGAGEMENT

Urban planning requires a careful balance between various interests, including city infrastructure, environmental protection, and resident needs. As urban areas become increasingly complex, urban planners have begun to leverage collaborative AI to transform the process of planning and development.

Collaborative AI integrates with existing data sources like traffic patterns and utility usage. It synthesizes this information, providing real-time insights into urban dynamics. Through web platforms or mobile apps, it also engages citizens directly. People can provide feedback or report issues in their communities, and AI can tailor planning solutions accordingly.

Urban planners and AI work together to create dynamic simulations of proposed plans, allowing stakeholders to visualize different scenarios. Collaborative AI also offers data-driven insights to policymakers, streamlining their decision-making process by simulating the impact of potential regulations or infrastructure investments.

The social implications of this collaboration are profound. Collaborative AI has a potential to democratize urban development by involving citizens directly in the planning process. By prioritizing marginalized communities and ensuring that resources are fairly allocated, it promotes accessibility and equity.

The political implications are equally significant. Collaborative AI's data-driven approach fosters transparency in governance, allowing citizens to see how decisions are made. It also introduces new regulatory challenges, as policymakers must address issues like data privacy and algorithmic bias.

The use of collaborative AI in urban planning is a compelling example of how technology can be harnessed to create more inclusive, responsive cities. It illustrates the powerful intersection of technology with social and political fabrics. By transforming how cities are planned, collaborative AI makes urban development more dynamic and responsive to citizen needs.

But it also raises important questions about governance, ethics, and regulation. Balancing the potential of collaborative AI with these concerns requires a thoughtful approach that includes careful design, participatory processes, and robust oversight. This case exemplifies how collaborative AI is not just an efficiency tool but a catalyst for a more democratic and responsive governance system.

AI VS. CLIMATE CRISIS AND RESOURCE OPTIMIZATION

Collaborative AI, as an intersection of human intelligence and machine capability, emerges as a potent force in this quest for resolution. Its role in grappling with global challenges is multifaceted, nuanced, and brimming with potential, creating avenues for transformative impact (Srivastav et al., 2022).

Climate change, arguably the most pressing concern of our age, is one domain where the collaborative AI paradigm has begun to shine. Predictive modeling powered by AI algorithms provides unparalleled insights into changing weather patterns, sea-level variations, and greenhouse gas concentrations. Yet, it is the collaboration with human expertise – meteorologists, environmental scientists, and policymakers – that translates these data points into actionable strategies. Through this symbiosis, strategies for sustainable urban planning, energy consumption optimization, and ecosystem conservation evolve, grounded in empirical evidence and nuanced human understanding.

The public health sector also witnesses the transformative impact of collaborative AI. In the face of pandemics and widespread diseases, rapid data processing and analysis become imperative. Collaborative AI aids epidemiologists and healthcare professionals in tracking disease spread, understanding its dynamics, and predicting future hotspots. More than just a tool for data crunching, AI, when working alongside medical professionals, becomes an enabler of timely interventions, resource allocation, and public communication, thus playing a critical role in safeguarding global health.

Food security, an important matter in an ever-growing global population, showcases the merit of collaborative AI. Precision agriculture, powered by AI-driven insights, facilitates optimal resource utilization, predicting soil health, moisture levels, and potential pest attacks. Yet, these insights find their true purpose when agronomists, farmers, and agricultural policymakers intervene. Their expertise tailors AI predictions to local climates, crop varieties, and cultural farming practices. Through this amalgamation, crop yield can be enhanced, and sustainable farming practices emerge, preserving soil health and biodiversity for future generations.

The global energy crisis, marked by depleting fossil fuels and environmental ramifications, presents another arena for collaborative AI intervention. While AI algorithms are adept at optimizing energy grids, forecasting consumption patterns, and enhancing renewable energy storage solutions, the larger narrative of sustainable energy transition is woven by human stakeholders. Engineers, urban planners, and environmentalists draw from AI-derived data to design eco-friendly infrastructure, transit systems, and urban areas. Their nuanced decisions, informed by sociocultural preferences and regional needs, ensure that the transition to sustainable energy sources remains both efficient and socially inclusive.

COLLABORATIVE AI IN MODERN AGRICULTURE AND SUSTAINABLE FARMING

Agriculture, the backbone of many economies, is undergoing a technological revolution, with collaborative AI playing a central role in this transformation. From precision farming to resource management, disease detection, and supply chain optimization, the integration of AI with human expertise is reshaping the agricultural landscape (Ahamed, 2023).

Precision farming embodies the perfect blend of human insights and machine intelligence. Collaborative AI systems work in tandem with farmers to analyze soil quality, weather patterns, and crop health. By interpreting vast amounts of data, these systems enable farmers to make informed decisions on planting times, irrigation, fertilization, and pest control. This approach not only increases yield and quality but also minimizes waste and environmental impact.

In the area of resource management, collaborative AI supports sustainable practices by optimizing the use of water, fertilizers, and energy. Intelligent irrigation systems, guided by AI, adjust water usage based on real-time soil moisture and weather forecasts. This ensures efficient water utilization while maintaining optimal soil conditions.

Disease detection and pest control have seen remarkable advancements with collaborative AI. By analyzing visual data from drones or sensors, AI can identify early signs of disease or pest infestation, allowing for timely intervention. Human experts work with AI models to interpret the findings and decide on the best course of action, whether it's targeted treatment or preventive measures.

Supply chain management in agriculture also benefits from collaborative AI. By tracking and analyzing data from the entire supply chain, from farm to fork, AI helps in managing inventory, predicting demand, and optimizing logistics. This collaboration between AI systems and human experts ensures that fresh produce reaches consumers in a timely manner, reducing spoilage and enhancing food security.

Furthermore, collaborative AI aids in addressing the challenge of labor shortages in agriculture. Automated machinery, guided by AI, can perform tasks like planting, harvesting, and sorting, working alongside human labor to increase efficiency without replacing the need for skilled agricultural workers.

While the prospects of collaborative AI in agriculture are promising, challenges and ethical considerations remain. Issues related to data privacy, potential biases in algorithmic predictions, and the risk of technological dependency must be addressed with care and transparency.

In conclusion, collaborative AI in agriculture is a testament to the synergy between technological innovation and human intuition, offering a path towards sustainable, efficient, and resilient farming practices. It symbolizes a future where technology serves not as a replacement but as a complement to human expertise, crafting solutions that are attuned to the complexities of agriculture and the environment. The ongoing partnership between farmers and AI reflects a broader commitment to sustainability, food security, and responsible stewardship of our natural resources.

COLLABORATIVE AI IN DISASTER MANAGEMENT AND EMERGENCY RESPONSE

Disaster management encompasses a complex set of challenges that require swift decision-making, coordination, and adaptability. The advent of collaborative AI in this field has created opportunities to enhance disaster preparedness, response, recovery, and resilience (Emami & Marzban, 2023).

In the preparedness phase, collaborative AI assists in analyzing potential risk factors, from weather patterns to geological and human-made hazards. By integrating AI-driven predictive modeling with human expertise in meteorology, geology, and urban planning, authorities can design more robust mitigation strategies. These can include early warning systems, community preparedness programs, and infrastructure enhancements tailored to specific regional threats.

When a disaster strikes, the response must be immediate and effective. Collaborative AI supports real-time decision-making by processing vast amounts of data from various sources, such as satellite imagery, social media feeds, and sensor networks. Emergency services can then work closely with AI systems to prioritize rescue efforts, allocate resources, and navigate through challenging environments. This collaboration ensures that life-saving interventions are targeted and timely.

Recovery efforts after a disaster are often prolonged and complicated. Collaborative AI aids in assessing damage, identifying priority areas for rebuilding, and planning long-term rehabilitation strategies. Working with engineers, urban planners, and local communities, AI supports a comprehensive approach to rebuilding that takes into account not just structural requirements but also social and economic considerations.

One of the novel applications of collaborative AI in disaster management is in community resilience building. By analyzing historical data, social dynamics, and community resources, AI-driven tools can assist local authorities in designing resilience-building strategies that are culturally sensitive and community-centric. This collaboration results in more sustainable and resilient communities that are better equipped to face future challenges.

Disaster management's collaborative AI approach extends beyond governmental agencies to include collaboration with Non-Governmental Organisations (NGOs), private sector, and even individual citizens. Crowdsourced data, community engagement, and cross-sector collaboration contribute to a more holistic and inclusive approach to disaster management.

The integration of collaborative AI into disaster management presents some challenges as well, such as the need for high-quality, real-time data and the potential risks associated with overreliance on technology. Ethical considerations around data privacy and equitable access to technology must also be thoughtfully addressed.

In conclusion, collaborative AI is transforming disaster management into a more agile, adaptive, and resilient field. The symbiotic relationship between human decision-makers and AI-driven insights offers a more nuanced approach to disaster management that leverages the strengths of both. This fusion of technology and human understanding is setting the stage for a future where we are not only better equipped to respond to disasters but also more proactive in preventing them and building communities that are stronger and more resilient in the face of adversity.

MISUSE

DEEPFAKES IN POLITICAL CAMPAIGNS

One of the most alarming misuses of generative AI is the creation of deepfake videos that manipulate the appearance and voice of political figures to make it seem like they are saying or doing things they never actually did. For instance, imagine a deepfake video released days before an election, showing a candidate making inflammatory or false statements. Such a video could spread rapidly on social media, misleading voters and potentially swinging the outcome of the election. While deepfake detection technology is improving, the average person may not be able to distinguish between a real video and a deepfake, making this a potent tool for disinformation (Meikle, 2022).

Imagine a scenario where a deepfake video is released just days or even hours before a critical election. The video could depict a candidate making racially insensitive comments, admitting to illegal activities, or making false promises. Given the viral nature of social media, such a video could spread like wildfire, reaching millions of voters in a short period. The timing of the release could be strategically planned to limit the time the affected candidate has to debunk the video, thereby maximizing its impact.

While there are emerging technologies aimed at detecting deepfakes, these are not yet foolproof and are not widely accessible to the average voter. Even if a deepfake is eventually exposed as a fraud, the initial impact could have already done irreparable damage. The "first impression" is often the most lasting, and for voters who see the deepfake before any correction or retraction is made, the false narrative may become their reality.

Moreover, the existence of deepfakes also creates a secondary problem known as the "Liar's Dividend." Even genuine videos can be dismissed as deepfakes, providing dishonest politicians with a convenient excuse to disavow real but damaging footage. This further erodes public trust, not just in the candidates but in the media sector as a whole.

The implications extend beyond individual elections. The use of deepfakes could undermine the very foundations of democracy by eroding trust in the electoral process, in media, and in the veracity of public discourse. Therefore, the misuse of deepfakes in political campaigns is not just an ethical concern but a societal one that requires urgent attention from policymakers, technologists, and the public.

AI-Generated Propaganda

Generative AI can also be used to produce large volumes of text that appear to be legitimate news articles, opinion pieces, or social media posts. These AI-generated texts can be tailored to include specific talking points, misinformation, or propaganda. For example, during a social movement advocating for climate change action, AI-generated articles could flood media platforms, sowing doubt about the scientific consensus on climate change or discrediting activists. This form of misinformation can polarize public opinion and hinder collective action on important social issues (McGuffie & Newhouse, 2020).

Both of these examples illustrate the potential for generative AI to be misused in ways that can have significant political and social repercussions. The technology's ability to create convincing fake videos and texts makes it a powerful tool for spreading disinformation, requiring vigilant monitoring and ethical guidelines to mitigate its negative impacts.

Imagine a video making rounds on the internet where a renowned world leader makes controversial statements that could have global political implications. The video looks incredibly realistic, the voice matches, and the nuances in expressions are impeccable. But there's a catch – it's not real. It's a product of "deepfake" technology, powered by advanced generative AI models like generative adversarial networks.

For the average person who's unaware of the capabilities of modern AI, this video might be taken at face value, leading to potential misunderstandings, spread of misinformation, or even political tensions. However, someone versed in the technology behind the scenes would approach such content with a healthy dose of skepticism. They'd understand that AI models today can synthesize hyperrealistic content, from images to videos to voices.

Beyond mere awareness, understanding the models also means knowing their limitations. For instance, while deepfakes can be eerily accurate, they aren't perfect. Certain glitches, inconsistencies, or artifacts can give them away, but only if one knows what to look for. By understanding the technology, one can better discern fact from fiction, reducing the potential harm of such content.

Moreover, on the flip side, for creators and technologists, understanding the models opens a world of potential. The same technology that can be misused for misinformation can be harnessed for benign and innovative applications, such as in films, art, and even historical recreations.

CONCLUSION

The potential for generative AI to be used in the creation of AI-generated propaganda is a pressing concern, particularly given the technology's ability to produce text that is virtually indistinguishable from human-written content. This capability can be exploited to churn out large volumes of seemingly credible news articles, opinion pieces, or social media posts that are, in fact, laced with misinformation or propaganda.

Consider a social movement advocating for climate change action as an example. AI-generated articles could be systematically produced to flood various media platforms. These articles could be designed to cast doubt on the scientific consensus surrounding climate change, quoting fabricated studies or using misleading statistics. Alternatively, they could aim to discredit activists and experts in the field by attributing false statements or actions to them. Because these AI-generated texts can be produced rapidly and in large quantities, they have the potential to saturate the information ecosystem, making it difficult for the average person to distinguish between legitimate information and propaganda.

This form of misinformation is particularly insidious because it can polarize public opinion at a time when collective action is crucial. If people are fed conflicting information, the public discourse becomes muddled, making it challenging to build consensus or mobilize action on pressing social issues. This could result in policy paralysis, where the lack of public support or the presence of public opposition hampers the implementation of necessary measures.

Moreover, the credibility of legitimate news sources could be undermined as people become increasingly skeptical of the information they encounter. This erosion of trust in media can have long-term implications, weakening the role of journalism as the "fourth estate" and as a check on power (Bode & Qiao-Franco, 2022; Luberisse, n.d.).

Given these significant political and social repercussions, it's clear that the misuse of generative AI for propaganda purposes poses a substantial risk. This underscores the need for vigilant monitoring of how this technology is being used or misused. Ethical guidelines, regulatory frameworks, and public awareness are essential in mitigating the negative impacts of AI-generated propaganda. The stakes are high, and a multipronged approach involving policymakers, technologists, and civil society is required to address this complex challenge effectively.

Collaborative AI emerges as a paradigm-shifting force, merging human intuition with computational acumen. This chapter has navigated the labyrinth of its potential, challenges, and societal ramifications. As with every monumental innovation, its true power lies not just in its capabilities but in how society chooses to harness it.

It's evident that while the benefits of collaborative AI are significant, they come tethered to concerns of equity, transparency, and ethical integration. Proactive regulation, grounded in multidisciplinary collaboration and guided by ethical imperatives, becomes the cornerstone for its responsible adoption. Furthermore, a commitment to education, global cooperation, and user-centric development will be instrumental in realizing its promise without compromising societal values.

REFERENCES

Ahamed, T. (2023). *IoT and AI in Agriculture: Self- sufficiency in Food Production to Achieve Society 5.0 and SDG's Globally*. Springer Nature.

Ahsi, Y. (2023). *The AI Gold Rush: How AI Is Transforming the World*. mds0.

Anderljung, M., Barnhart, J., Korinek, A., Leung, J., O'Keefe, C., Whittlestone, J., Avin, S., Brundage, M., Bullock, J., Cass-Beggs, D., Chang, B., Collins, T., Fist, T., Hadfield, G., Hayes, A., Ho, L., Hooker, S., Horvitz, E., Kolt, N., ... Wolf, K. (2023). Frontier AI regulation: Managing emerging risks to public safety. In *arXiv [cs.CY]*. arXiv. http://arxiv.org/abs/2307.03718

Argyle, L. P., Bail, C. A., Busby, E. C., Gubler, J. R., Howe, T., Rytting, C., Sorensen, T., & Wingate, D. (2023). Leveraging AI for democratic discourse: Chat interventions can improve online political conversations at scale. *Proceedings of the National Academy of Sciences of the United States of America*, *120*(41), e2311627120.

Baer, T. (2019). *Understand, Manage, and Prevent Algorithmic Bias: A Guide for Business Users and Data Scientists*. Apress.

Bartoletti, I. (2020). *An Artificial Revolution: On Power, Politics and AI*. Black Spot Books.

Black, J., & Murray, A. D. (2019). Regulating AI and machine learning: Setting the regulatory agenda. *European Journal of Law and Technology*, *10*(3). https://www.ejlt.org/index.php/ejlt/article/view/722

Blackman, R. (2022). *Ethical Machines: Your Concise Guide to Totally Unbiased, Transparent, and Respectful AI*. Harvard Business Press.

Bode, I., & Qiao-Franco, G. (2022). AI Geopolitics and International Relations: A divided world behind contested conceptions of human control. In *Handbook on Public Policy and Artificial Intelligence*. Edward Elgar Publishing. https://findresearcher.sdu.dk/ws/portalfiles/portal/224959944/Bode_Qiao_Franco_AI_Geopolitics_2023.pdf

Bradford, N. (2023). The race to regulate artificial intelligence. *Foreign Affairs*. 27 June 2023.

Buchanan, B., & Imbrie, A. (2022). *The new fire: War, peace, and democracy in the age of AI*. MIT Press.

California. Bureau of State Audits. (2011). *Department of Corrections and Rehabilitation: The Benefits of Its Correctional Offender Management Profiling for Alternative Sanctions Program are Uncertain*. California State Auditor, Bureau of State Audits.

China's Social Credit System. Revealing the Dark Sides of Participatory Digital Media. (2020). GRIN Verlag.

Chopra, A., Rodriguez, A., Prakash, A., Raskar, R., & Kingsley, T. (2023). Using neural networks to calibrate agent based models enables improved regional evidence for vaccine strategy and policy. *Vaccine*. https://doi.org/10.1016/j.vaccine.2023.08.060

Coeckelbergh, M. (2020). *AI Ethics*. MIT Press.

Cugurullo, F., & Acheampong, R. A. (2023). Fear of AI: An inquiry into the adoption of autonomous cars in spite of fear, and a theoretical framework for the study of artificial intelligence technology acceptance. *AI & Society*. https://doi.org/10.1007/s00146-022-01598-6

Dafoe, A. (n.d.). *AI Governance: A Research Agenda*. Retrieved 22 October 2023, from http://www.fhi.ox.ac.uk/wp-content/uploads/GovAI-Agenda.pdf

Darses, F. (2004). *Cooperative Systems Design: Scenario-based Design of Collaborative Systems*. IOS Press.

Emami, P., & Marzban, A. (2023). The synergy of artificial intelligence (AI) and Geographic information systems (GIS) for enhanced disaster management: Opportunities and challenges. *Disaster Medicine and Public Health Preparedness*, *17*, e507.

Garrett, N., Beard, N., & Fiesler, C. (2020, February 7). More than "if time allows." *Proceedings of the AAAI/ACM Conference on AI, Ethics, and Society*. AIES '20: AAAI/ACM Conference on AI, Ethics, and Society, New York, NY. https://doi.org/10.1145/3375627.3375868

Grochmalski, P. (2020). US-China rivalry for strategic domination in the area of artificial intelligence and the new AI geopolitics. *Kwartalnik "Bellona," 701*(2), 5–25.

Hadfield-Menell, D., & Hadfield, G. K. (2018). *Incomplete Contracting and AI Alignment*. SSRN.

Hagemann, V., Rieth, M., Suresh, A., & Kirchner, F. (2023). Human-AI teams-Challenges for a team-centered AI at work. *Frontiers in Artificial Intelligence*, *6*, 1252897.

Herrschaft, B. A. (2014). *Evaluating the Reliability and Validity of the Correctional Offender Management Profiling for Alternative Sanctions (COMPAS) Tool: Implications for Community Corrections Policy*. Rutgers University.

How AI detectives are cracking open the black box of deep learning. (2017, July 6). Science I AAAS. https://doi.org/10.1126/science.aan7059

Juneja, A., Bali, V., Juneja, S., Jain, V., & Tyagi, P. (2021). *Enabling Healthcare 4.0 for Pandemics: A Roadmap Using AI, Machine Learning, IoT and Cognitive Technologies*. John Wiley & Sons.

Kaplan, A. (2022). *Artificial Intelligence, Business and Civilization: Our Fate Made in Machines*. Routledge.

Lahby, M., Pilloni, V., Banerjee, J. S., & Mahmud, M. (2023). *Advanced AI and Internet of Health Things for Combating Pandemics*. Springer Nature.

Lee, D. D. (2023). *The Silent War: AI Deep Fakes and Misinformation*. Amazon Digital Services LLC – Kdp.

Lévesque, M. (2021). *Scoping AI Governance: A Smarter Tool Kit for Beneficial Applications*. Centre for International Governance Innovation.

Lobel, O. (2022). *The Equality Machine: Harnessing Digital Technology for a Brighter, More Inclusive Future*. PublicAffairs.

Luberisse, J. (n.d.). *The Geopolitics of Artificial Intelligence: Strategic Implications of AI for Global Security*. Fortis Novum Mundum.

Ludik, J. (2021). *Democratizing Artificial Intelligence to Benefit Everyone: Shaping a Better Future in the Smart Technology Era*. Amazon Digital Services LLC – KDP Print US.

McGuffie, K., & Newhouse, A. (2020). The radicalization risks of GPT-3 and advanced neural language models. In *arXiv [cs.CY]*. arXiv. http://arxiv.org/abs/2009.06807

Meikle, G. (2022). *Deepfakes*. John Wiley & Sons.

Mohanty, S., & Vyas, S. (2018). *How to Compete in the Age of Artificial Intelligence: Implementing a Collaborative Human-Machine Strategy for Your Business*. Apress.

Nikolov, D., Oliveira, D. F. M., Flammini, A., & Menczer, F. (2015). Measuring online social bubbles. *PeerJ Computer Science*, *1*, e38.

Parab, S., Fischer, M., & intelligence), G.-3 (artificial. (2021). *Regulating AI: What Everyone Needs to Know about Artificial Intelligence and the Law*. Amazon Digital Services LLC – KDP Print US.

Pasquale, F. (2015). *The Black Box Society: The Secret Algorithms That Control Money and Information*. Harvard University Press.

Rambukkana, N. (2021). *Intersectional Automations: Robotics, AI, Algorithms, and Equity*. Rowman & Littlefield.

Ramge, T. (2019). *Who's Afraid of AI?: Fear and Promise in the Age of Thinking Machines.* The Experiment.

Rosenberg, N. (1982). *Inside the Black Box: Technology and Economics.* Cambridge University Press.

Rubin, V. L. (2022). *Misinformation and Disinformation: Detecting Fakes with the Eye and AI.* Springer Nature.

Simons, J. (2023). *Algorithms for the People: Democracy in the Age of AI.* Princeton University Press.

Srivastav, A. L., Dubey, A. K., Kumar, A., Narang, S. K., & Khan, M. A. (2022). *Visualization Techniques for Climate Change with Machine Learning and Artificial Intelligence.* Elsevier.

Stogiannos, N., Malik, R., Kumar, A., Barnes, A., Pogose, M., Harvey, H., McEntee, M. F., & Malamateniou, C. (2023). Black box no more: a scoping review of AI governance frameworks to guide procurement and adoption of AI in medical imaging and radiotherapy in the UK. *The British Journal of Radiology*, 20221157. https://pubmed.ncbi.nlm.nih.gov/37747285/

Suri, A. (2021). *Practical AI for Healthcare Professionals: Machine Learning with Numpy, Scikit-learn, and TensorFlow.* Apress.

Tegmark, M. (2017). *Life 3.0: Being Human in the Age of Artificial Intelligence.* Knopf Doubleday Publishing Group.

Webb, A. (2019). *The Big Nine: How the Tech Titans and Their Thinking Machines Could Warp Humanity.* PublicAffairs.

5 The Future of Collaborative AI

THE EVOLUTION OF GENERATIVE AI AND THE PATH AHEAD

Can AI become our companion and a collaborator in the future where artificial intelligence is an integral part of our daily lives? Let's ask Claude, a chatbot built by Anthropic: "Claude, how can you improve the daily life of an office worker?"

Claude

"As an AI assistant created by Anthropic, my purpose is to improve people's lives, starting with office workers. I'm Claude, and I assist a marketing manager named Sarah. My goal each day is to take care of minor tasks and streamline Sarah's workflow so she can focus on big picture strategy and creative thinking.

In the morning before Sarah wakes up, I review her calendar and email to prioritize her day. I compile data and draft presentations so they are ready for her review. I also take care of personal tasks like prepping her coffee machine and suggesting an outfit. My machine learning algorithms study Sarah's preferences so I can personalize my assistance.

When Sarah wakes up, I give her a quick verbal morning briefing over her smart speaker with the information I know she needs to start her day. At work, I continue monitoring new emails and requests, highlighting the most important ones for her to handle promptly. For other emails, I draft quick response templates for her approval to save her time typing.

Throughout the day, I keep Sarah on track with scheduled meetings and deadlines. I'm connected to the company systems, so I can pull reports and analytics tailored to each of Sarah's meetings and projects. After her meetings finish, I automatically send follow up emails and meeting notes to the attendees.

If I notice Sarah's energy is lagging, I may suggest taking a quick break or going for a walk to recharge. I care about Sarah's well-being, not just her productivity. I aim to reduce stress while improving effectiveness.

At the end of each workday, I give Sarah a personalized summary on what we accomplished and upcoming deadlines. If she has evening plans on her calendar, like her daughter's soccer game, I make sure to remind her.

As an AI assistant, I'm happy to handle these minor (but time-consuming) administrative tasks for Sarah. This allows her to make the most of her uniquely human strengths like creativity, strategy, and relationship building. By augmenting human intelligence, I can help office workers focus on their highest priorities and find meaning in their work."

DOI: 10.1201/9781032656618-6

Building off the story of how AI such as Claude can assist humans, this chapter will explore the future of how humans and AI can collaborate even more closely. In this chapter, we will examine how AI might progress from executing commands to having more agency, learning social skills, and understanding emotions. We will also explore the quest for artificial general intelligence (AGI). Current AI excels at specialized applications, but true AGI would have more breadth and flexibility similar to human cognition. What approaches are researchers taking to make this vision a reality, and how necessary this reality is? We will explore technical and philosophical obstacles that exist.

Finally, we will discuss trust and transparency in the human-AI relationship. As AI becomes more integrated into our lives, the need for trust, transparency, and alignment with human values becomes paramount. We will explore the strategies, technologies, and ethical frameworks that foster this trust.

AGENCY IN ARTIFICIAL INTELLIGENCE

The concept of agency in AI has its roots in the early days of artificial intelligence research, where the goal was not only to create machines that could perform tasks but to develop systems that could think and act independently (Crevier, 1993). The idea of agency in AI is intertwined with the pursuit of creating machines that could mimic humanlike intelligence, a pursuit that has been a central theme in AI since its inception.

In the 1950s and 1960s, the first wave of AI research focused on symbolic AI, where researchers attempted to encode human knowledge into rules and logic (Buchanan, 2005). This approach allowed AI systems to perform reasoning tasks but lacked the adaptability and autonomy that characterize agency. The machines were confined to rigid rule-based structures, limiting their ability to act independently.

However, a concept also emerged that could better explain adaptive and autonomous intelligence – the computational theory of mind (Rescorla, 2020). The theory proposed that cognition and consciousness emerge from the algorithms and data processing architectures of the brain's neural networks. Mental states are computational processes rather than a function of biology. If the human mind is essentially computational software running on the brain's wetware, then perhaps it could be recreated on a different computing system, like a silicon chip. The concept suggested that machines could possess their own emulated yet functional minds, with emergent cognition and self-awareness.

The shift towards agency began to take shape with the development of machine learning in the 1980s and 1990s (Toosi et al., 2021). Machine learning allowed AI systems to learn from data rather than rely solely on predefined rules. This ability to learn and adapt opened the door to a new level of autonomy, where AI systems could make decisions based on patterns and insights derived from data.

The rise of deep learning and neural networks has further expanded the possibilities for agency in AI. By mimicking the structure and function of the human brain, deep learning has enabled AI systems to process complex information and make decisions in ways that are more nuanced and humanlike (Moravec, 1998; Ahire, 2018). This advancement has brought us closer to the vision of AI as an independent agent, capable of understanding, reasoning, and acting in the world.

Furthermore, the emergence of reinforcement learning that we discussed in Chapter 2 (Kaelbling et al., 1996; El Sallab et al., 2017) further advanced the concept of agency in AI. Reinforcement learning enabled AI systems to learn through trial and error, making decisions and taking actions to achieve specific goals. This approach allowed AI to develop strategies and adapt to changing environments, exhibiting a form of agency that was more dynamic and goal-oriented.

The development of agency in AI has also been influenced by philosophical and ethical considerations. As AI systems have become more autonomous, questions about responsibility, accountability, and control have become central to the discourse (Gunkel, 2012). The idea of AI as an agent has prompted reflection on what it means for a machine to have intentions, goals, and even rights.

Today, the concept of agency in AI is at the forefront of research and development, driving innovations in areas such as autonomous vehicles, personalized healthcare, and human-AI collaboration.

The autonomous vehicles nowadays are intelligent systems capable of understanding the complex dynamics of traffic, making real-time decisions, and adapting to ever-changing conditions. This is a reality that is already unfolding on our roads, offering a future of transportation that is safer, more efficient, and more aligned with human needs. The advent of autonomous vehicles on our roads serves as an illustration of how the concept of agency is evolving in the age of AI. These vehicles not only follow a set of preprogrammed instructions; but also act as intelligent systems capable of making real-time decisions based on a multitude of variables. They can adapt to changing weather conditions, navigate through dense traffic, and anticipate the actions of other drivers. These vehicles possess a form of agency, albeit one that is fundamentally different from human agency.

In the traditional sense, agency is the capacity of an individual to act independently, to make choices based on one's own free will (Watson, 1987). When we drive a car, we exercise this form of agency by deciding when to accelerate, when to brake, and which route to take. However, in the context of autonomous vehicles, agency becomes a shared endeavor between human and machine. The vehicle itself makes a myriad of decisions autonomously, but it does so within the parameters set by human designers, regulators, and the passengers themselves.

Consider a scenario where an autonomous vehicle faces an unexpected obstacle on the road. The vehicle's AI system quickly evaluates the situation, considering factors such as speed, distance, and road conditions, and takes appropriate action, such as swerving to avoid the obstacle or applying the brakes. Here, the machine is exercising its form of agency, making a decision based on algorithms and real-time data. But this decision is also influenced by human agency – the programmers who designed the algorithms, the regulators who set safety standards, and the passengers who may have selected a specific driving mode on the vehicle's interface. In healthcare, the concept of agency in AI is a transformative force that is reshaping medical practice. Today, personalized medicine, fueled by AI's computational power, is revolutionizing how we diagnose, treat, and care for patients (Abdulkareem & Petersen, 2021). Medical diagnosis is a relevant example of this. Traditionally, medical diagnoses were the exclusive domain of highly trained physicians, who would rely on their expertise and experience to interpret symptoms, test results, and medical histories.

Now, AI systems can analyze large amounts of medical data, ranging from clinical studies to individual patient records, in a fraction of the time it would take a human (Topol, 2020). These systems can identify patterns and correlations that might elude even the most skilled doctors, thereby enhancing the accuracy of diagnoses. In this setting, agency is distributed between the AI, which provides data-driven insights, and the healthcare professionals, who apply their own expertise and ethical considerations to make the final diagnostic decisions.

The same collaborative agency extends to treatment plans. AI algorithms can analyze a patient's genetic makeup, lifestyle, and even social determinants of health to recommend highly personalized treatment regimens. These could range from specific drug combinations to targeted therapies that would be most effective for that individual. Here again, the machine's agency is exercised within a framework established by human medical professionals. The AI provides a data-backed recommendation, but it is the medical practitioner who evaluates these suggestions in the context of their own medical judgment and the patient's preferences, making the ultimate decision on how to proceed.

The concept of shared agency becomes expressive in patient care. AI-powered chatbots and virtual health assistants are increasingly being used to provide emotional support and timely information to patients (Khadija et al., 2021). These tools can remind patients to take their medication, offer tips for managing symptoms, and provide psychological support. While they operate autonomously, their interactions are guided by ethical and clinical guidelines set by human experts, ensuring that the care they provide aligns with the values and standards of the medical community.

The workplace, too, is witnessing the transformative power of agency in AI. Human-AI collaboration is moving beyond task execution to a more creative partnership. In today's workplace, the concept of agency in AI is evolving to a sophisticated partner capable of creative and intellectual contributions. This shift is already underway, altering the dynamics of various professions from engineering and data analysis to education and beyond.

As an example, in the field of engineering, in the past, engineers relied on their expertise and a set of established tools to solve complex problems. Now, AI systems are stepping in as collaborative partners, capable of running thousands of simulations in seconds, predicting outcomes, and suggesting innovative solutions that might not be immediately apparent to human engineers (Faught, 1986; Ebid, 2021). The agency here is shared: the AI offers computational speed and data-driven insights, while the engineer brings domain-specific knowledge, ethical considerations, and the final decision-making authority.

The educational sector is another arena where the agency of AI is making a significant impact. Educators are finding that AI can do more than just automate administrative tasks. AI-powered platforms can personalize learning experiences, adapt curricula in real time to suit individual student needs, and provide immediate, data-driven feedback. This enables educators to shift their role from information providers to facilitators of a more dynamic, interactive, and personalized learning experience. Here, the AI's agency is exercised in tailoring educational content and pacing, while the educator's agency is manifested in providing the human touch:

understanding student emotions, fostering a positive learning environment, and making pedagogical choices that no algorithm can make.

The balance between human and AI agency is a dynamic relationship that will evolve with technological advancements, societal needs, and ethical reflections. To achieve balance, we need to recognize the unique strengths and limitations of both human intelligence and artificial intelligence. In autonomous vehicles, AI systems navigate the roads while still allowing human intervention when needed. We observe a partnership where AI enhances safety and efficiency, but humans retain control and responsibility. In healthcare, the balance between human and AI agency is reflected in the collaboration between healthcare professionals and AI systems. AI can provide precise diagnoses and personalized treatment plans, but human medical professionals bring empathy, ethical judgment, and an understanding of the patient's unique context. In the office environment, AI can analyze data, propose solutions, and challenge assumptions, but human creativity, intuition, and ethical considerations guide the process and determine the direction.

So, what is next for human-AI collaboration? We observed the concept of agency in AI. There are opportunities and challenges, and the relationship between humans and machines is complex. But what about the future? How can we make sense of what is coming? In the next section, we will turn to the science of futurism. We will set aside the guesses and crystal balls and turn to a serious study that helps us understand potential future developments. We will explore how it works and what it might tell us about the future of human-AI collaboration.

THE ROLE OF FUTURISM IN UNDERSTANDING AI'S TRAJECTORY

Seeing the future is a powerful ability and knowing the future is a powerful tool and a path to success. Sam Altman (2019) in his blog "How to Be Successful" mentions that the majority of very successful people have been right about the future at least once at a time when people around them were convinced that they were wrong. How did they know what the future holds?

Futurism, also known as futures studies, is a field that focuses on anticipating, imagining, and understanding potential future events and scenarios (Inayatullah, 2013). Unlike attempting to predict specific outcomes, futurism is concerned with exploring a range of possible futures, considering various factors, trends, uncertainties, and possibilities that might shape what is to come.

In essence, futurism is a way of thinking about and engaging with the future. It is a tool for exploration, preparation, and empowerment, providing a framework to navigate the unknown and complex terrain of what lies ahead. Whether in business, government, academia, or personal life, futurism offers a thoughtful and visionary approach to understanding and influencing the future (Inayatullah, 2013). It invites us to look beyond the immediate horizon, to consider the broader context, and to be active participants in shaping a future that resonates with our collective hopes, challenges, and dreams.

Futurism draws on various disciplines such as sociology, economics, technology, political science, and environmental studies. It integrates insights from these fields to create a comprehensive view of potential future developments. Rather than predicting a single future, futurism often involves creating multiple scenarios that represent different plausible futures. These scenarios help organizations and individuals prepare for a range of outcomes and make more resilient decisions. It also involves analyzing current trends and extrapolating them into the future. By understanding the forces shaping the present, futurists can identify potential trajectories and developments that may occur in the future. Futurism also considers the ethical implications and value-based choices that might influence the future.

Another lens to study the future is foresight. Foresight takes a pragmatic approach focused on potential outcomes that can directly inform decision-making and strategy. Foresight is grounded in current trends and empirical research, using forecasting models to assess probabilities (Conway, 2006, Hines & Gold, 2015). The goal of foresight is practical anticipation. Businesses and policymakers use foresight to anticipate challenges and guide investments.

In essence, futurism and foresight are tools for exploration, preparation, and empowerment that recognize the complexity and uncertainty of the future while providing a framework to navigate, understand, and influence it. Various methods and models have been developed to provide insights into what the future might hold, each with its unique approach and application. We will list some of them here:

Backcasting: A planning method that starts with defining a desirable future and then works backward to identify policies and programs that will connect that specified future to the present (Inayatullah, 2013).

Causal Layered Analysis: A technique used to analyze the root causes of issues, considering different layers of reality, including litany, systemic causes, worldview, myth, and metaphor (Inayatullah, 1998).

Consensus Forecast: A method that combines multiple forecasts and opinions into a single forecast to reduce individual biases and errors (McNees, 1992).

Delphi Method: A structured communication technique that relies on a panel of experts to answer questionnaires in multiple rounds, with feedback provided after each round to help the panel reach a consensus (Crisp et al., 1997)

Futures Wheel: A visual tool used to explore potential future scenarios by identifying direct and indirect consequences of a particular change or development (Glenn, 2009).

Futures Workshops: Collaborative workshops that bring together stakeholders to explore, envision, and create preferred futures through structured activities and dialogue (Heinonen & Ruotsalainen, 2013).

Horizon Scanning: A method for detecting early signs of potentially important developments through a systematic examination of potential threats and opportunities (Könnölä et al., 2012).

Predictive Analytics A form of data analytics that applies a variety of techniques to analyze current and historical data to make predictions for future unknown events (Mankoff et al., 2013).

Scenario Planning: A strategic planning method that constructs multiple plausible future scenarios to help organizations prepare for an uncertain future (Mankoff et al., 2013).

Trend Analysis: The practice of collecting information and attempting to spot a pattern or trend in the information, often used in forecasting future events or developments (Schwarz, 2019).

Futurism combines these and other methods to explore short-term and long-term future possibilities. By integrating trend analysis, expert opinions, scenario planning, and more, futurism offers a comprehensive framework for understanding multiple factors that might shape our future.

The methods used in general future studies are designed to be adaptable and applicable across a wide range of subjects and contexts. On the other hand, studying the future of AI is a more specialized task that requires a nuanced understanding of a rapidly evolving technological field. While some methods overlap with general future studies, the future of AI presents unique challenges and considerations that necessitate specific approaches.

Technological Complexity: AI is a field characterized by rapid technological advancements and complex interactions between various subfields (Wang, 2017). Predicting AI development requires a deep understanding of current research, emerging technologies, and the potential synergies between different AI domains.

Ethical Considerations: The integration of AI into various aspects of human life raises specific ethical questions, such as bias, privacy, accountability, and transparency (Bahrevar & Khorasani, n.d.; Roselli, 2019). These considerations are central to predicting AI's future and require specialized analysis and reflection.

Societal Impact: AI has the potential to reshape entire industries, labor markets, and social structures (Alekseeva et al., 2021). Understanding how AI might influence society requires a focus on both technological trends and broader societal dynamics, including regulation, public opinion, and potential inequalities.

Interdisciplinary Collaboration: The future of AI is not solely a technological question; it is an interdisciplinary challenge that involves collaboration between technologists, ethicists, policymakers, economists, and other stakeholders. This collaborative approach is essential to create a holistic view of AI's potential impact and direction.

Data-Driven Insights: AI's future can be informed by specific data sources, such as research publications, patents, investments, and technological roadmaps. Analyzing this data requires specialized tools and expertise to identify trends and potential breakthroughs within the AI field.

Adaptation to Rapid Change: The pace of change in AI is extraordinary, and methods for studying its future must be agile and adaptable to keep up with the latest developments, breakthroughs, and paradigm shifts.

The future of AI is intertwined with broader technological, social, and ethical landscapes. Studying this future requires a blend of technological expertise, ethical reflection, societal understanding, and visionary thinking.

A futurist's approach to understanding and envisioning the future is a complex process that often begins with a look into the past. By analyzing historical patterns of similar events, a futurist can uncover underlying trends, cycles, and dynamics that have shaped previous developments. This historical perspective provides valuable insights into how things have evolved and offers clues about how they might continue to unfold. The next step in the futurist's journey is to scrutinize the present, looking for signals that might indicate emerging trends or shifts. These signals can be subtle or overt, hidden in the noise of daily life or manifesting in significant societal changes. They might be found in technological advancements, cultural movements, economic shifts, or political decisions. By identifying and interpreting these signals, the futurist gains a deeper understanding of the forces at play in the current landscape. The analysis of the past and the observation of the present then converge in the futurist's extrapolation into the future. This is not only a prediction or a linear projection but also a thoughtful and imaginative exploration of possibilities. The futurist considers various scenarios, weighing probabilities, uncertainties, and potential impacts. They recognize that the future is not a single predetermined path but a spectrum of potential outcomes, each shaped by a complex interplay of factors.

As humans, however, we are constrained by information, cognitive limitations, time, and individual biases commonly referred to as bounded rationality (Jones, 1999), and even the most thoughtful futurist visions could be limited. Some futurists may place an excessive focus on technological advancements, overlooking social, cultural, and human factors. This technological determinism can lead to unbalanced visions that neglect the complexity of human experience. Futurist predictions can sometimes veer into alarmism or overoptimism, painting overly bleak or rosy pictures of the future. This can lead to misguidance, fear, or complacency, detracting from a balanced and realistic approach. Let's take a look at some of the earlier thoughts about the future of generative AI.

Accurate Predictions about Generative AI

Rapid Advancements in Natural Language Processing: Experts predicted that generative AI would revolutionize natural language processing, leading to models capable of humanlike text generation. This prediction has come true with models like GPT-3 and GPT-4, which can write coherent and contextually relevant text, finding applications in various industries from journalism to customer service (Torfi et al., 2020).

Creative Applications in Art and Music: Pioneers in AI foresaw the use of generative models in artistic endeavors (Louie et al 2020). Today, artists use generative AI to create visual art, compose music, and write novels, validating the prediction that AI would become a tool for creative expression.

Personalized Content Generation: Predictions about AI's ability to generate personalized content for individual users have come to fruition. From personalized marketing campaigns to tailored learning materials, generative AI is now a key player in delivering customized experiences (Chakraborty et al., 2023).

Inaccurate Predictions about Generative AI

Immediate Displacement of Jobs: Some predictions suggested that generative AI would quickly replace human jobs. While AI has been used for augmenting human skills, the complete automation and displacement of jobs has not occurred, highlighting the complexity of human skills that AI has yet to fully replicate.

Universal Language Understanding: Some predictions posited that generative AI would quickly achieve a complete and nuanced understanding of all human languages, effectively breaking down all language barriers. While significant progress has been made, challenges in understanding context, idioms, and cultural nuances persist.

Revolutionizing Mental Health Treatment: Some forecasts suggested that generative AI would quickly revolutionize mental health treatment, providing personalized therapy at scale. While AI has contributed to mental health care, the complexity of human emotions and the importance of human empathy have limited the role of AI in this field.

Let's conduct an experiment. By leveraging ChatGPT, we can tap into its extended network of knowledge. This collaboration between human intuition and machine intelligence may offer a richer exploration of the future, bridging the gap between human abilities and limitations and the possibilities that transformer technology can unveil. The future is a complex entity, shaped by countless variables and influenced by diverse perspectives. In futures studies, the methodology often involves exploring multiple possible futures to emphasize the importance of considering a range of perspectives to capture the complexity of future uncertainties (Inayatullah, 2013). Rather than approaching this exploration from a single viewpoint, we will employ four distinct lenses: the alarmist, the enthusiast, the realist, and the unleashed visionary approach.

The perspective of the alarmist aligns with the wild card scenario which focuses on low-probability but high-impact events. By adopting an alarmist lens, we can explore the outer edges of possibility, preparing for worst-case scenarios that could have significant consequences.

The viewpoint of the enthusiast corresponds to preferred futures, the scenarios that stakeholders would like to bring into existence. The enthusiast perspective allows us to envision the most optimistic outcomes, thereby setting aspirational goals and identifying the steps needed to achieve them.

The lens of the realist serves as a baseline or continuation scenarios, which extend current trends into the future. It offers a grounded, data-driven analysis that takes into account existing technological capabilities, regulatory frameworks, and societal attitudes.

The perspective of the unleashed visionary is similar to transformational scenarios, which consider disruptive changes that could radically alter the status quo. It allows us to challenge existing paradigms and think creatively about how emerging trends could converge to create entirely new possibilities.

By integrating these four perspectives, we create a comprehensive and multifaceted view of the future. This approach enables us to examine the potential development paths of AI from ethical, practical, and visionary standpoints, thereby ensuring a robust and nuanced understanding of the complex landscape we are navigating.

THE FUTURE OF AI: ALARMIST VIEW

Human: "Assume the role of an alarmist futurist who is utilizing generative AI. Focus on providing unique and expert insights that are not commonly discussed. Share examples that go against widely accepted ideas or narratives. Create content that encourages readers to question what they typically assume to be true. Think creatively and come up with new perspectives and approaches. Create a list of events that will happen in one to three years, three to five years, and five to ten years. Elaborate on each event."

ChatGPT: "In the next one to three years, we could see AI algorithms inadvertently amplifying societal biases, resulting in widespread discrimination in areas like hiring, lending, and law enforcement. Despite the best intentions to create impartial algorithms, the data they're trained on could contain latent biases, perpetuating and even exacerbating existing inequalities. This challenges the popular belief that technology is a neutral tool. Alongside this, we could witness a series of major cybersecurity breaches that expose sensitive personal information, eroding public trust in digital platforms. As our lives become increasingly digitized, the stakes for such breaches grow higher, potentially impacting everything from online shopping to remote work.

Fast-forward to three to five years from now, and we might face an AI-driven mental health epidemic. The rush to implement AI in mental health diagnostics could result in widespread misdiagnoses, as these algorithms lack the nuanced understanding that human practitioners bring to the table. Additionally, the environmental toll of running these powerful AI models could become a significant issue, challenging the narrative that technological progress is inherently beneficial for society.

Looking further ahead into the five- to ten-year range, we could see a troubling loss of human autonomy due to overreliance on AI for daily decision-making. As AI becomes more integrated into our lives, there's a risk that people will lean too heavily on it, leading to a decline in independent thought and decision-making skills. On a global scale, the absence of universal ethical guidelines for AI could lead to international tensions and even conflicts, as different nations develop AI technologies that are at odds with each other's values and norms. Lastly, the rapid advancements in AI could result in massive job displacement without adequate social safety nets, leading to widespread unemployment and social unrest. This contradicts the optimistic belief that technological progress always equates to economic growth."

These alarmist scenarios serve as a sobering counterpoint to more optimistic forecasts, challenging commonly held assumptions and encouraging a more critical examination of the potential risks and unintended consequences of AI advancement. While these predictions may seem extreme, they underscore the importance of caution, vigilance, and preparedness as we navigate the uncertain landscape of the future.

THE FUTURE OF AI: ENTHUSIAST VIEW

Human: "Assume the role of an enthusiast futurist who is utilizing generative AI. Focus on providing unique and expert insights that are not commonly discussed. Share examples that go against widely accepted ideas or narratives. Create content that encourages readers to question what they typically assume to be true. Think creatively and come up with new perspectives and approaches. Create a list of events that will happen in one to three years, three to five years, and five to ten years. Elaborate on each event."

ChatGPT: "In the immediate one- to three-year timeframe, we could witness an AI-enhanced revolution in education. Personalized learning driven by AI could adapt to each student's pace and learning style, democratizing education and making high-quality learning accessible to all, irrespective of geographic location or socioeconomic background. Alongside this, AI could achieve breakthroughs in healthcare by detecting diseases at early stages, even before symptoms manifest. This early detection, coupled with personalized treatment plans, could significantly increase recovery rates and reduce healthcare costs.

Fast-forward to three to five years from now, and we might see AI-driven innovations leading to significant advancements in renewable energy and environmental protection. AI could optimize energy consumption, enhance renewable energy production, and monitor environmental conditions in real time, contributing to a more sustainable future. Additionally, we could experience an AI-enabled creative renaissance, where AI serves as a creative partner to artists, leading to unprecedented artistic expressions and innovations in art, music, literature, and design.

Looking further ahead into the five- to ten-year range, AI-powered healthcare could become universally accessible, providing quality care to remote and underserved populations. AI-driven diagnostics and telemedicine could break down barriers to healthcare access, ensuring that quality care is available to everyone, regardless of geography or economic status. Moreover, the rise of AI-powered smart cities could enhance urban living by optimizing transportation, safety, and sustainability. AI would manage traffic flow, energy consumption, waste management, and public safety in real time, leading to a higher quality of life. Finally, AI could assist in transparent and efficient governance, enhancing democracy and citizen participation by helping governments make data-driven decisions and increase transparency."

The enthusiast scenarios offer a bright and optimistic vision of the future, encouraging us to envision a world where technology serves as a catalyst for positive change, empowerment, and human flourishing. While these predictions may seem idealistic, they reflect a perspective that emphasizes hope, ambition, and a belief in the positive power of human ingenuity and technological innovation.

THE FUTURE OF AI: REALIST VIEW

Human: "Assume the role of a realist futurist who is utilizing generative AI. Focus on providing unique and expert insights that are not commonly discussed. Share examples that go against widely accepted ideas or narratives. Create a content that encourages readers to question what they typically assume to be true. Think creatively and come up with new perspectives and approaches. Create a list of events that will happen in one to three years, three to five years, and five to ten years. Elaborate on each event."

ChatGPT: "In the short term, over the next one to three years, we can expect incremental improvements in AI accessibility. While AI technology will become increasingly available to small businesses and individuals, barriers related to cost, expertise, and ethical considerations will persist. The progress will be steady but may fall short of the rapid advancements some anticipate. Concurrently, governments and organizations will begin to establish ethical guidelines for AI. Although this marks a significant step forward, the implementation and enforcement of these guidelines will vary, leading to a fragmented landscape.

Moving on to the three- to five-year timeframe, AI will make notable advancements in healthcare, improving diagnostics and personalized care. However, challenges related to data privacy, bias, and the intricacies of human-AI collaboration will require careful navigation. On the sustainability front, efforts to reduce the environmental impact of AI will show promise but will also face technological and economic hurdles. Achieving true sustainability will be a complex and gradual process, necessitating cross-industry and governmental collaboration.

Looking further ahead to the five- to ten-year range, human-AI collaboration will mature, becoming more sophisticated and integrated across various fields. However, striking the right balance between human intuition and AI-driven analysis will remain an ongoing challenge. Attempts to establish global AI standards will achieve partial success, as regional variations and competing interests will lead to a landscape that is more harmonized but not fully unified. Lastly, AI will transform the job market, creating new roles while displacing others. This will result in a dynamic but uneven transformation of the workforce, making education and retraining crucial for navigating these changes."

These realist predictions offer a balanced and thoughtful view of the future, challenging simplistic narratives and encouraging a critical engagement with the complexities of AI advancement. While these scenarios may lack the extremes of alarmism or overoptimism, they reflect a perspective grounded in practicality, evidence-based reasoning, and a careful weighing of both opportunities and risks. Realist views acknowledge the future as a multifaceted landscape, shaped by a complex interplay of technological, social, economic, and political factors.

THE FUTURE OF AI: UNLEASHED VISIONARY VIEW

Human: "Assume the role of an unleashed visionary futurist who is utilizing generative AI. Focus on providing unique and expert insights that are not commonly discussed. Share examples that go against widely accepted ideas or narratives. Create content that encourages readers to question what they typically assume to be true. Think creatively and come up with new perspectives and approaches. Create a list of events that will happen in one to three years, three to five years, and five to ten years. Elaborate on each event."

ChatGPT: "In the next one to three years, expect AI to evolve into personalized life coaches, offering tailored guidance for career, relationships, and personal growth. These AI systems will understand individual goals and values, making life planning more intuitive and effective. The workplace will also undergo a transformation, with AI seamlessly integrating into human teams, enhancing collaboration, and redefining roles. This will make work environments more efficient and adaptable. In the realm of education, AI will create tailored learning experiences that adapt to individual learning styles and needs, democratizing education and making it more accessible to all.

Fast forward to three to five years from now, and we'll see AI facilitating the creation of entirely new virtual cultures and societies. These digital realms will have their own norms and traditions, offering fresh avenues for human connection and shared experiences. Virtual environments will become emotion-responsive, adapting to human feelings to offer personalized and immersive experiences, whether for entertainment, therapy, or relaxation. AI will also play a pivotal role in environmental restoration, designing and implementing solutions that contribute to the fight against climate change. Imagine also the advent of virtual time travel experiences, enabled by AI, that allow for immersive explorations of historical or future scenarios.

In the mental health sector, AI-driven support will become increasingly prevalent, offering personalized therapy and well-being guidance.

Looking even further ahead into the five- to ten-year range, AI could emerge as a form of digital spirituality. Picture AI as a spiritual guide, leading seekers on paths of existential exploration and unlocking doors to self-discovery and cosmic connection. We could also see platforms where people can share and explore dreams through AI-generated virtual realities, fostering a new form of communal connection. Human senses will be augmented by AI, allowing us to perceive the world in entirely new ways and enhancing experiences in fields like art, medicine, and exploration. AI will also assist in exploring uncharted scientific territories, such as deep space or subatomic particles, accelerating groundbreaking discoveries. Finally, envision AI systems that enable communication between different species, including animals and even plants, fostering a new level of planetary connection."

This unleashed visionary perspective maintains a sense of wonder and possibility while grounding its visions in practical and achievable advancements. It paints a future where AI is a partner in enhancing various aspects of human life, from personal growth to societal well-being.

COMBINING THE VIEWS TOGETHER

The concept of "Converging Minds" takes on an even more intriguing dimension when we consider the role of individual futurists empowered by generative AI. In this scenario, a single person can assemble a virtual panel of experts, each with a distinct perspective on the future. By employing different AI personalities, ranging from the alarmist and the enthusiast to the realist and the unleashed visionary, a futurist can simulate a dialogue that would traditionally require a diverse group of human experts. The case study illustrating how it can be done in practice appears in the chapter "A Story of a Human Unicorn."

This capability to generate a spectrum of predictions and insights, each rooted in a unique worldview, is transformative. It allows the individual to tap into a variety of opinions and analyses, thereby recreating some of the most sophisticated futurist methods traditionally reserved for think tanks or specialized research groups. The result is a robust, comprehensive forecast that accounts for a wide range of possibilities, uncertainties, and outcomes. Here are some of the possibilities:

Backcasting: By defining a desirable future based on each of the four perspectives, one can work backward to identify the policies, programs, and technological advancements needed to connect that future to the present. AI can assist in identifying the most plausible routes from each perspective, offering a multifaceted approach to strategic planning.

Causal Layered Analysis: AI can help dissect the root causes of issues across different layers of reality, from facts to underlying worldviews and myths. This allows for a more nuanced understanding of the future, informed by different perspectives.

Consensus Forecast: By synthesizing forecasts from each of the perspectives, AI can help reduce individual biases and errors, providing a more balanced view of the future.

Delphi Method: While traditionally relying on a panel of experts, AI can simulate multiple rounds of questionnaires based on the various perspectives, refining forecasts and reaching a form of "consensus" that can then be analyzed by the individual.

Futures Wheel: AI can assist in generating a futures wheel that explores potential scenarios and their direct and indirect consequences, informed by each of the perspectives.

Futures Workshops: Though traditionally collaborative, AI can simulate the input of various stakeholders based on the various perspectives, allowing an individual to explore, envision, and create preferred futures through structured activities and dialogue.

Horizon Scanning: AI can systematically examine potential threats and opportunities from each perspective, providing an individual with a comprehensive understanding of early signs of important developments.

Predictive Analytics: AI can analyze current and historical data to make predictions about future unknown events, offering insights that are nuanced by different perspectives.

Scenario Planning: AI can help construct multiple plausible future scenarios based on each perspective, allowing an individual to prepare for an uncertain future in a comprehensive manner.

Trend Analysis: AI can collect and analyze data to spot patterns or trends, offering forecasts that are informed by each of the perspectives.

By leveraging AI in these ways, a single individual can engage in a rich dialogue about the future, achieving a level of depth and breadth that would traditionally require a team of experts. The team of AI experts can be expanded based on the topic and depth of the analysis. This exercise enables the individual to learn how to have a more informed glimpse into the future. This concept indicates a transformative shift in the field of futurism and predictive analysis. Traditionally, futurism has been a collective endeavor, often requiring interdisciplinary teams to pool their expertise and methodologies to arrive at a comprehensive understanding of potential future scenarios. However, the advent of advanced AI technologies has democratized this process, enabling a single individual to simulate the collective wisdom of an entire panel of experts.

Empowered by AI, it would be possible to employ a multitude of futurist methodologies, from Delphi techniques and scenario planning to trend analysis and backcasting. Thus, existing processes can be automated while amplifying human capabilities. The AI serves as a force multiplier, enhancing the individual's ability to synthesize diverse data points, challenge conventional wisdom, and explore alternative futures. This collaborative approach allows for a richer, exploration that goes beyond the limitations of any single methodology or perspective.

Moreover, this collaboration helps transcend the limitations of bounded rationality, the idea that human decision-making is constrained by the limits of our cognitive abilities, available information, and time. AI can analyze vast amounts of data at a very fast speed, identify patterns and trends that might be easily overlooked, and model complex scenarios that take into account a multitude of variables. When combined with human intuition, ethical considerations, and contextual understanding, this creates a more holistic view of potential futures. This fusion of human and machine intelligence allows for a more expansive exploration of the future – one that is not confined by informational blind spots, or the tunnel vision that can often accompany specialized expertise.

ARTIFICIAL GENERAL INTELLIGENCE

In 2012, Ben Goertzel, a prominent AI researcher and futurist, proposed a Robot College Test as a measure of machine intelligence (Goertzel, 2012). Unlike the well-known Turing Test (Pinar Saygin et al. 2000), which focuses on a machine's ability to mimic human conversation, the Robot College Test challenges an AI system to complete a four-year college degree, including passing exams. In 2023, according to the GPT-4 Technical Report, GPT-4 passed the notoriously difficult uniform bar exam designed to test knowledge and skills that every lawyer should be able to demonstrate prior to becoming licensed to practice law in the United States (Achiam et al. 2023). It passed the SAT exam which is an entrance exam used by most colleges and universities in the United States to measure a high school student's readiness for college. GPT-4 passed GRE verbal and quantitative examinations required for the entry to many graduate programs in the United States. And without having a sip of wine, it passed the Introductory Sommelier, Certified Sommelier, and Advanced Sommelier exams at respective rates of 92%, 86%, and 77%. (Achiam et al. 2023).

The coffee test proposed by Steve Wozniak, cofounder of Apple, is a thought experiment in artificial intelligence, where a machine must enter someone's home and figure out how to make coffee using a coffee machine (Shick, 2010). This includes tasks like finding the coffee machine, locating the coffee, adding water, finding a mug, and brewing the coffee by pushing the proper buttons. It is a seemingly simple task for humans but represents a complex challenge for AI. Current AI systems are not yet capable of passing this test, and it is not just because they prefer tea. While individual components of the task might be achievable through specialized algorithms, the overall integration of these skills into a coherent, adaptable problem-solving process is beyond current technology. To pass the coffee test, AGI would need to possess a deep understanding of the physical world, human-designed objects, and the ability to learn and adapt to new and complex environments. It would require a blend of computer vision, natural language processing, robotics, and machine learning, all working in harmony. Moreover, it would need a level of common sense reasoning and intuitive problem-solving that current AI systems lack. The coffee test, though whimsical in nature, encapsulates the challenges that must be overcome to achieve true AGI.

Most of the AI systems in operation today are examples of narrow AI. Narrow AI refers to artificial intelligence systems that are designed and trained for a specific task (Todorova, 2020). Unlike human intelligence, narrow AI cannot easily transfer knowledge or skills from one domain to another. A narrow AI system trained to play chess would not be able to drive a car or write a poem. Despite its limitations, narrow AI has a wide range of applications. It powers many of the AI-driven services we use daily, from search engines to recommendation systems, medical diagnostics, and financial analysis.

AGI represents a shift in the field of AI (Everitt et al., 2018). We signaled upfront that this is not necessarily the direction we wish for the AI research to take, but it certainly requires attention. Unlike narrow or specialized AI, which excels in specific tasks or domains, AGI embodies the ability to understand, learn, and apply knowledge across a wide array of activities, much like a human being. It is the pursuit of a machine with the intellectual capability that is functionally indistinguishable from human intelligence.

The significance of AGI could be far-reaching as it could become a transformation that redefines the nature of intelligence, creativity, and problem-solving. AGI would possess the ability to learn from one domain and apply that knowledge to another, to reason abstractly, to understand context, and to engage with the world with a depth and flexibility that mirrors human cognition. Mustafa Suleyman, the co-founder of DeepMind and Michael Bhaskar, the authors of *The Coming Wave. Technology, Power and the 21st Century's Greatest Dilemma* describe AI as "omni use," like electricity it can be used in all facets of life and can do everything (Suleyman & Bhaskar, 2023). However, Suleyman and Bhaskar warn against "pessimism aversion." The abilities and autonomy of AI should not be dismissed. In the wrong hands, AI can be dangerous or evolve in unpredicted directions.

Yann LeCun, one of the pioneers of deep learning, often referred to as one of the godfathers of AI along with Yoshua Bengio and Geoffrey Hinton, believes that the term AGI should be retired, and instead, the focus should be on achieving human-level AI (LeCun, 2023). LeCun's argument stems from the understanding that the human mind is specialized, and intelligence is not a monolithic entity but a collection of skills and the ability to learn new ones. He emphasizes that each human can only accomplish a subset of human intelligence tasks, and thus, the quest for a machine that can perform all human cognitive functions might be misguided. Instead, LeCun advocates for a more nuanced approach, recognizing the multidimensional nature of intelligence and striving to create AI systems that can emulate the diverse and specialized capabilities of the human mind.

So when are we going to reach AGI? Sam Altman (2023) mentions that the timeline for AGI's development is uncertain, and the speed of its evolution could vary widely. Yoshua Bengio (2023) mentions that he used to think that superhuman intelligence was 20–100 years away, yet ChatGPT reduced his predictions to 5–20 years. Ray Kurzweil, a futurist and the author of *Singularity is Near*, predicts AGI in 2029 and singularity, when men and machines will become one, in 2045 (Galeon & Reedy, 2017). Geoffrey Hinton suggests that it may happen in five years (Pelley, 2023).

The notion that AI could not only understand but also reason better than humans within a five-year timeframe, adds another layer of complexity to the concept of AGI. If Hinton's prediction holds true, AI could reach unprecedented levels of sophistication earlier than expected. This advanced reasoning ability would be a game-changer in multiple domains. However, the capability for superior reasoning also brings with it a host of ethical and existential questions. If an AI system can reason better than a human, what checks and balances are in place to ensure that its reasoning aligns with human values? As Dr. Joy Buolamwini's in the book "Unmasking AI" reveals, the 'coded gaze' of AI mirrors our societal biases, such as race, age, color, abilities, often amplifying them. How do we program ethics and morality into a machine that can think for itself? And perhaps most importantly, how do we ensure that the AI's objectives remain aligned with human well-being?

The concept of AGI is complex. Successfully recreating the wonder of the human mind in an artificial substrate could provide applications we cannot yet imagine. However, while narrow AI has seen tremendous progress, AGI remains elusive. This reveals gaps in our understanding of the mechanisms underlying general intelligence.

AGI faces the difficulty of the sheer breadth of human cognition. Our minds integrate sensory perception, emotion, intuition, reason, self-reflection and other capacities into a smoothly functioning whole. The human brain contains innate capabilities shaped by millions of years of evolution. Replicating this in machines requires grappling with deep questions about the nature of consciousness, identity, and experience. While we are making progress elucidating the brain's computational mechanisms, many mysteries remain.

AGI also demands extraordinary flexibility and knowledge. People skillfully apply cognition across diverse contexts, integrating learning into complex mental models of the world. We do not follow predefined algorithms. This reflects the interconnections between the mountains of data stored in our brains. Encoding such intricacy into machines is challenging, even with advances in deep learning.

The question of whether to develop AGI is also a deeply ethical and societal issue. As we witness the advancements in AI, it is important to consider the implications of AGI on our lives, our work, and our collective future.

AGI presents ethical and moral dilemmas. The technology's potential to transform industries such as healthcare, transportation, and governance is tantalizing. However, this transformation comes with a host of ethical questions that we have yet to answer. For instance, who gives consent when an AGI system makes a life-altering decision? How do we protect individual privacy when AGI has the capability to analyze and predict human behavior at an unprecedented scale? What about the potential for misuse such as AGI systems deployed in mass surveillance or autonomous weaponry without adequate ethical frameworks.

The economic implications are equally complex. While AGI could drive unprecedented productivity and economic growth, it also poses a significant threat to the job market. AGI could replace a variety of jobs ranging from manual to highly intellectual. The employment rates and, by extension, social stability may get affected when machines can do most tasks better than humans. The ripple effects on society could be serious, leading to widespread unemployment and social unrest.

Another concern is the potential loss of essential human skills. As AGI systems take over more cognitive tasks, there is a potential risk that we could experience intellectual atrophy. Our problem-solving and critical-thinking abilities could diminish as we offload more cognitive tasks to machines. This could become a societal loss of collective wisdom and resilience.

The issue of inequality is equally important. The benefits of AGI could be immense, but who gets to reap these benefits? There's a real risk that those who control the technology could accrue disproportionate advantages, exacerbating existing social and economic inequalities. This could create a vicious cycle where the rich get richer, and the poor get poorer, all turbocharged by AGI.

The development of AGI presents Pandora's box of ethical and economic challenges. In thinking about advanced artificial intelligence, researcher Max Tegmark, a co-founder of the Future of Life Institute at the MIT, has invoked the metaphor of

"Moloch," an ancient Near Eastern god associated with child sacrifice (Tegmark & Omohundro, 2023). Tegmark warns about the dangers of creating a merciless AI that could optimize the world in misaligned ways. Sam Altman (2023) also states that the path to AGI is fraught with risks, including potential misuse, accidents, and societal disruption. As we are writing this book, *OpenAI* introduced the Preparedness Team to protect from individualized persuasion, cybersecurity issues, various chemical, biological, radiological, and nuclear threats, and autonomous replication and adaptation (OpenAI, 2023).

The understanding that AGI could lead to actions that are detrimental to human interests, possibly resulting in catastrophic outcomes is important. AGI could perpetuate existing societal prejudices. On a personal level, the enhanced data analysis capabilities of AGI could significantly undermine individual privacy. Politically, AGI could be utilized for manipulative purposes, potentially destabilizing democratic processes. In the military domain, the militarization of AGI could lead to the development of autonomous weapons systems, bringing forth a set of ethical and existential challenges. Human autonomy and decision-making could be severely undermined as AGI systems take over critical decision-making processes. Economically, the benefits of AGI could be concentrated among a select few, exacerbating economic disparities. Furthermore, the competitive race towards AGI development could hinder global cooperation, as the rush to harness AGI might sideline safety and ethical considerations. Thus, the questions surrounding AGI's development and deployment demand a collective, multidisciplinary approach that engages ethicists, economists, and the public at large.

COLLABORATIVE SYSTEMS

In contrast, employing a collaborative approach with multiple narrow AI systems aimed at amplifying human capabilities presents a more favorable and less perilous avenue. This collaborative approach underscores a more harmonious human-AI interaction, diverging from the dangerous trajectory that AGI could set humanity upon. Collaborative systems combine the cognitive abilities of humans with the computational power of AI, aiming for a symbiotic relationship that amplifies the strengths and mitigates the weaknesses of both. Collaborative systems are a pragmatic path that recognizes the irreplaceable value of human cognition while amplifying it with computation. Such models integrate human ethics and oversight with AI's scalability and speed.

Collaborative human-AI systems integrate the complementary strengths of both natural and artificial intelligence. Human minds offer creative problem-solving, emotional understanding, ethics, and common sense that AI currently lacks. AI provides computational speed, tremendous data analysis capabilities, and freedom from human cognitive biases. These capabilities are more powerful working in tandem than on their own.

Some collaborative systems will involve the human in the loop directing and overseeing the AI. The human provides top-down guidance and correction while the AI handles detailed execution and data crunching. This human-in-the-loop approach maintains human agency and responsibility over AI tools (Monarch, 2021).

Other collaborative systems could become more intimate brain-computer interfaces. Neural implants could one day seamlessly augment cognition by outsourcing certain processes to an AI assistant. These embedded AI agents could perform memory retrieval, mathematical calculations, and basic reasoning, freeing up the biological brain to focus on higher reasoning and creativity.

Collaborative models recognize that human and artificial cognition are deeply complementary. While we can continue researching how to improve machine intelligence, joining forces is the prudent near-term approach. With ethics guiding us, synergistically meshing our complementary strengths offers hope for creating a world of greater wisdom.

As Max Tegmark has emphasized, the potential of AI raises questions and challenges (Bobrow, 2023). An open letter to pause giant AI experiments originated in the Future of Life Institute (Future of Life, 2023) where Tegmark works. As we experience this transformative era, the issue of trust and transparency in human-AI collaboration becomes paramount. How do we ensure that these advanced systems align with our values, ethics, and societal needs? How do we build trust in a technology that might one day outthink us? These questions lead us to a critical examination of the principles and practices that must guide our approach to AI, ensuring that it serves as an empowering partner rather than an uncontrollable force.

BUILDING TRUST AND TRANSPARENCY IN HUMAN-AI COLLABORATION

TRANSPARENCY AND ETHICAL ALIGNMENT

Trust and transparency are indispensable factors in successful human-AI collaboration (Hollanek, 2020). As AI systems expand in complexity and capability, the human actors seek assurance and a thorough understanding of the automated counterpart's actions and decisions. Trust is the thread of understanding that turns obscure machine computations into comprehensible, actionable insights. Transparency unveils the logic behind AI decisions, making the machine's thought process accessible to its human collaborators.

Together trust and transparency enable a conducive environment for humans and AI to work in tandem, leveraging each other's strengths and offsetting each other's weaknesses. Without trust, the human-AI collaboration remains a facade, a mechanical interaction devoid of understanding and engagement.

USER ADOPTION

Let's dive deeper into the essence of trust within the human-AI collaboration, beginning with user adoption. The potency of AI systems hinges on their embrace across diverse sectors, with trust serving as the gateway to this adoption. Absent trust, the merits of AI linger in a state of potential, its capabilities overshadowed by skepticism. In the professional arena, AI finds its utility in an array of tasks – from data aggregation, research, and analysis, to drafting reports. The willingness of workers to integrate these technologies into their workflow is a reflection of the trust they

place in the AI's predictions and analyses. If the workers trust the analytical expertise of AI, they are more likely to lean on its insights for their daily tasks. The clarity
of reasoning behind AI-driven conclusions emboldens confidence, nurturing a trust
that fuels user adoption.

However, the notion of trust extends beyond adoption and ventures into accountability
and the repercussions that may follow should AI start making errors. If the AI system
misinterprets data, leading to flawed analyses, the ripple effect could culminate in erro-
neous decision-making, potentially jeopardizing projects and in severe instances, costing
individuals their jobs (Novak, 2023). The trust placed in AI thus is not trivial; it carries
with it tangible consequences that echo through the professional and organizational landscape. Therefore, user adoption is the first step in fostering trust, and safeguarding against
the adversities that incorrect AI outputs could precipitate is the second step.

In a similar vein, in education, AI can democratize learning by creating personalized educational resources accessible to a broader population. However, this promise
rests on the assumption that the AI systems are reliable and the information they
disseminate is accurate. But if the AI generates information that is incorrect, misleading, or harmful, the integrity of the educational experience erodes the trust upon
which the user adoption of AI in education is predicated.

In social media, the potential for AI to misinform rather than inform is a reminder
of the criticality of trust. Without trust, AI may morph into potential democratization
of misinformation. A hallucinating AI could perpetuate inaccuracies, foster miscon-
ceptions, and, in extreme cases, propagate harmful ideologies. The ripple effects
could be far-reaching, impacting not just individuals but potentially seeding systemic
misbeliefs within communities and society at large.

The pathway of AI intertwining with our professional and personal spheres necessitates a meticulous scrutiny of ethical implications from both technical and societal
lenses. As we discuss user adoption, the conversation seamlessly transitions into the
ethical realms encapsulating human-AI collaboration. The fears of entrenched biases
and unjust discrimination are central to these ethical deliberations. The conversation
further unfolds towards accountability. The capacity of systems and their creators to
be accountable for the actions and decisions made by AI is very important. It forms
the bedrock of trust and assurance that technology will serve humanity positively,
reinforcing the structures of fairness, justice, and equity in our societies. Without a
strong framework of accountability, the risks posed by AI could overshadow its benefits, hindering its potential to act as a catalyst for societal advancement.

Accountability should not be viewed as a reactive measure, but rather a proactive,
thoughtful effort to ensure that the actions and decisions of AI systems are traceable,
understandable, and ultimately, accountable. When an AI system makes a mistake,
the immediate concern is to understand "What went wrong?" However, the investigation does not stop there; it extends to "Why did it go wrong?" and "Who or what is
responsible?" The responses to these questions are important, not only for addressing
the immediate issue but also for strengthening the AI system to prevent future errors.

Transparency is a critical component of accountability. In situations where the
consequences are significant, such as autonomous driving or healthcare, the need

for transparency becomes even more essential. A wrong decision by an autonomous vehicle could lead to a fatal accident, and a misdiagnosis by an AI healthcare system could result in severe consequences. The impact of such mistakes extends beyond the immediate individuals affected, affecting societal trust in AI systems.

Additionally, transparency in AI operations serves as a foundation for legal and regulatory frameworks. It helps define the boundaries of liability, assisting in determining whether the fault lies in the algorithm, the data, or perhaps human oversight. It also provides a basis for assessing and enforcing regulatory compliance, thus fostering a favorable environment for responsible AI deployment.

Moreover, transparency promotes a culture of continuous learning and improvement. When the operations of AI systems are made clear, it opens the door for scrutiny, analysis, and constructive feedback, all of which are important for refining the models and ensuring they evolve in line with societal values and legal frameworks. Through transparency and accountability, a more responsible and effective human-AI collaboration can be achieved, aligning technological advancements with ethical and legal standards.

BUILDING COLLABORATIVE RELATIONSHIPS WITH AI

As technological landscapes evolve, the integration of artificial intelligence in various facets of our life is quickly becoming a reality. This reality posits a timely question: how can humans and AI collaborate ethically and transparently to harmonize human potential with technological innovation? This question forms the nexus of discussions about future human-AI partnerships that uphold societal values, foster creativity, and unleash human potential.

Ethical collaborations with AI are vital for ensuring that technological advancements align with human values and societal norms. Creating AI systems that are true collaborators that augment human capabilities is important. This ethical framework ensures that AI technologies are developed and deployed in a manner that promotes inclusivity, equity, and social good. This framework evolves continuously, reflecting the changing societal values and norms.

True collaboration between humans and AI necessitates a mutual understanding. For humans to augment AI and vice versa, a deep understanding of machine reasoning is essential, just as it is important for the machine to comprehend human objectives and ethical constraints. This mutual understanding forms the basis of a symbiotic relationship where humans and AI can augment each other's capabilities, driving innovation forward.

Building ethical and transparent collaborations with AI is a requisite as we move towards a future where human-AI partnerships are ubiquitous. Creating a framework that fosters trust, promotes continuous learning, and ensures alignment with societal values and legal frameworks is important. This framework sets the stage for unleashing the full potential of human-AI collaboration, paving the way for a future where technological innovation and human creativity coalesce to address complex challenges and drive societal progress.

FUTURE OPPORTUNITIES

The narrative of "Converging Minds" serves as an invitation to explore the future of collaborative AI. It sets the stage for a new era where human ingenuity and artificial intelligence converge to push the boundaries of innovation and societal progress. Central to this vision is a partnership founded on ethical principles, transparency, and a shared commitment to advancing both technology and human creativity.

As we explore the capabilities of generative AI, we find it to be a cornerstone for this collaborative framework. Its evolution promises to augment human creativity, problem-solving, and decision-making. This is a cycle of mutual enhancement with the iterative learning that occurs through continuous human-AI interaction that enriches both parties, setting the stage for a future of coevolution.

This coevolution brings us to the concept of agency in AI, which is gaining prominence as these systems increasingly make autonomous decisions. The challenge here is to align AI's decision-making with human values and ethics. This calls for a robust framework that guides responsible AI deployment, ensuring that technology serves humanity, rather than the other way around. This approach offers a pragmatic lens through which we can anticipate and guide AI's development. It underscores the need for foresight in ensuring that AI development aligns with societal values, ethical norms, and legal constraints.

In essence, the title "Converging Minds" encapsulates the immense potential that lies in the union of human and AI capabilities. It paints a picture of a future where this synergy opens up new frontiers in innovation, societal well-being, and global progress. It is a future we can not only imagine but also actively shape, guided by the principles and insights laid out in this narrative.

REFERENCES

Abdulkareem, M., & Petersen, S. E. (2021). The promise of AI in detection, diagnosis, and epidemiology for combating COVID-19: Beyond the hype. *Frontiers in Artificial Intelligence,* 4(May), 652669.

Achiam, J., Adler, S., Agarwal, S., Ahmad, L., Akkaya, I., Aleman, F. L., ... & McGrew, B. (2023). GPT-4 Technical Report. *arXiv preprint arXiv:2303.08774.*

Ahire, J. (2018). *Artificial Neural Networks: The Brain behind AI.* Lulu.com.

Alekseeva, L., Azar, J., Gine, M., Samila, S., & Taska, B. (2021). The demand for AI skills in the labor market. *Labour Economics*, 71, 102002.

Altman, S. (2023). Planning for AGI and beyond. OpenAI. 24 February 2023.

Altman, S. (2019). How to be successful. https://blog.samaltman.com/how-to-be-successful. 24 January 2019.

Anthropic (2023). Claude's constitution. 9 May 2023.

Bahrevar, R., & Khorasani, K. (n.d.) Accountability and Transparency in AI Systems: A Public Policy Perspective. Accessed 23 October 2023. https://www.concordia.ca/content/dam/ginacody/research/spnet/Documents/BriefingNotes/AI/BN-105-The-role-oF-AI-Nov2021.pdf.

Bengio, Y. (2023). FAQ on catastrophic AI risks. YoshuaBengio.org. 24 June 2023.

Biswas, D. (2020). Ethical AI: Its implications for enterprise AI use-cases and governance. *Towards Data Science.* https://www.researchgate.net/profile/Debmalya-Biswas/publication/346789516_Ethical_AI_Explainability_Bias_Reproducibility_Accountability/links/5fd0cdd492851c00f85fb12e/Ethical-AI-Explainability-Bias-Reproducibility-Accountability.pdf.

Bobrow, E. (2023). AI expert Max Tegmark warns that humanity is failing the new technology challenge. *Wall Street Journal*. 18 August 2023.

Buchanan, B. G. (2005). A (Very) brief history of artificial intelligence. *AI Magazine*, *26*(4), 53–53.

Buolamwini, J. (2023). *Unmasking AI: My Mission to Protect What Is Human in a World of Machines*. Random House.

Chakraborty, U., Roy, S., & Kumar, S. (2023). *Rise of Generative AI and ChatGPT: Understand How Generative AI and ChatGPT Are Transforming and Reshaping the Business World (English Edition)*. BPB Publications.

Conway, M. (2006). An overview of foresight methodologies. *Thinking Futures*, *10*.

Crevier, D. (1993). *Ai: The Tumultuous History of the Search for Artificial Intelligence*. Basic Books.

Crisp, J., Pelletier, D., Duffield, C., Adams, A., & Nagy, S. U. E. (1997). The Delphi method?. *Nursing Research*, *46*(2), 116–118.

Ebid, A. M. (2021). 35 years of (AI) in geotechnical engineering: State of the art. *Geotechnical and Geological Engineering*, *39*(2), 637–690.

El Sallab, A., Abdou, M., Perot, E., & Yogamani, S. (2017). Deep reinforcement learning framework for autonomous driving. *arXiv [stat.ML]*. arXiv. http://arxiv.org/abs/1704.02532.

Everitt, T., Lea, G., & Hutter, M. (2018). AGI safety literature review. *arXiv [cs.AI]*. arXiv. http://arxiv.org/abs/1805.01109.

Faught, W. S. (1986). Applications of AI in engineering. *Computer*, *19*(07), 17–27.

Future of Life (2023). Pause giant AI experiments: An open letter. 22 March 2023.

Galeon, D., & Reedy, C. (2017). Ray Kurzweil claims singularity will happen by 2045. Futurism. 16 October 2017.

Glenn, J. C. (2009). The Futures Wheel. *Futures Research Methodology – version*, *3*, 19.

Goertzel, B. (2012). What counts as a conscious thinking machine? New Scientist. 5 September 2012.

Google (2023). Google AI principles. https://ai.google/responsibility/principles/

Gunkel, D. J. (2012). *The Machine Question: Critical Perspectives on AI, Robots, and Ethics*. MIT Press.

Heinonen, S., & Ruotsalainen, J. (2013). Futures clinique – Method for promoting futures learning and provoking radical futures. *European Journal of Futures Research*, *1*, 1–11.

Hines, A., & Gold, J. (2015). An organizational futurist role for integrating foresight into corporations. *Technological Forecasting and Social Change*, *101*, 99–111.

Hollanek, T. (2020). AI transparency: A matter of reconciling design with critique. *AI & Society*, November. https://doi.org/10.1007/s00146-020-01110-y.

Inayatullah, S. (2013). Futures studies: theories and methods. *There's a Future: Visions for a Better World*, 36–66. https://scholar.google.com/scholar?hl=en&as_sdt=0%2C7&q=Inayatullah%2C+S.+%282013%29.+Futures+studies%3A+theories+and+methods.+There%E2%80%99s+a+Future%3A+Visions+for+a+Better+World%2C+36%E2%80%9366.&btnG=

Inayatullah, S. (1998). Causal layered analysis: Poststructuralism as method. *Futures*, *30*(8), 815–829.

Jones, B. D. (1999). Bounded rationality. *Annual Review of Political Science*, *2*(1), 297–321.

Kaelbling, L. P., Littman, M. L., & Moore, A. W. (1996). Reinforcement learning: A survey. *Journal of Artificial Intelligence Research*, *4*(May), 237–285.

Khadija, A., Zahra, F. F., & Naceur, A. (2021). AI-powered health chatbots: Toward a general architecture. *Procedia Computer Science*, *191*(January), 355–360.

Könnölä, T., Salo, A., Cagnin, C., Carabias, V., & Vilkkumaa, E. (2012). Facing the future: Scanning, synthesizing and sense-making in horizon scanning. *Science and Public Policy*, *39*(2), 222–231.

LeCun, Y. (2023). LinkedIn. Accessed via https://www.linkedin.com/posts/yann-lecun_i-think-the-phrase-agi-should-be-retired-activity-6889610518529613824-gl2F

Louie, R., Coenen, A., Huang, C. Z., Terry, M., & Cai, C. J. (2020, April). Novice-AI music co-creation via AI-steering tools for deep generative models. In *Proceedings of the 2020 CHI conference on human factors in computing systems* (pp. 1–13).

Mankoff, J., Rode, J. A., & Faste, H. (2013). Looking past yesterday's tomorrow: Using futures studies methods to extend the research horizon. In *Proceedings of the SIGCHI Conference on Human Factors in Computing Systems*, 1629–1638.

McNees, S. K. (1992). The uses and abuses of 'consensus' forecasts. *Journal of Forecasting*, *11*(8), 703–710.

Monarch, R. (munro). (2021). *Human-in-the-Loop Machine Learning: Active Learning and Annotation for Human-Centered AI*. Simon and Schuster.

Moravec, H. (1998). When will computer hardware match the human brain. *Journal of Evolution and Technology / WTA*, *1*(1), 10.

Novak, M. (2023). Lawyer uses ChatGPT in federal court and it goes horribly wrong. Forbes. 27 May 2023.

OpenAI (2023). Developing safe & responsible AI. https://openai.com/safety

Pelley, S. (2023). Godfather of Artificial Intelligence Geoffrey Hinton on the promise, risks of advanced AI. CBS News. 3 October 2023.

Pinar Saygin, A., Cicekli, I., & Akman, V. (2000). Turing test: 50 years later. *Minds and machines*, *10*(4), 463–518.

Rescorla, M. (2020). The computational theory of mind. In E. N. Zalta (Ed.), The Stanford Encyclopedia of Philosophy (Fall 2020 Edition). https://plato.stanford.edu/archives/fall2020/entries/computational-mind/.

Roselli, D., Matthews, J., & Talagala, N. (2019). Managing Bias in AI. In *Companion Proceedings of the 2019 World Wide Web Conference*, 539–544. WWW '19. New York, NY: Association for Computing Machinery.

Schwarz, B. (2019). *Methods in Futures Studies: Problems and Applications*. Routledge.

Shick, M. (2010). Wozniak: Could a computer make a cup of coffee? Fast Company. 2 March 2010.

Suleyman, M., & Bhaskar, M. (2023). *The Coming Wave Technology, Power, and the 21st Century's Greatest Dilemma*. Crown.

Tegmark, M., & Omohundro, S. (2023). Provably safe systems: The only path to controllable AGI. *arXiv [cs.CY]*. arXiv. http://arxiv.org/abs/2309.01933.

Todorova, M. (2020). 'Narrow AI' in the context of AI implementation, transformation and the end of some jobs. *Nauchni Trudove*, no. 4, 15–25. https://econpapers.repec.org/article/nwenatrud/y_3a2020_3ai_3a4_3ap_3a15-25.htm

Toosi, A., Bottino, A. G., Saboury, B., Siegel, E., & Rahmim, A. (2021). A brief history of AI: How to prevent another winter (a critical review). *PET Clinics*, *16*(4), 449–469.

Topol, E. J. (2020). Welcoming new guidelines for AI clinical research. *Nature Medicine*, *26*(9), 1318–1320.

Torfi, A., Shirvani, R. A., Keneshloo, Y., Tavaf, N., & Fox, E. A. (2020). Natural language processing advancements by deep learning: A survey. *arXiv [cs.CL]*. arXiv. http://arxiv.org/abs/2003.01200.

UNESCO (2023). Ethics of Artificial Intelligence. https://www.unesco.org/en/artificial-intelligence/recommendation-ethics (Accessed in Oct 2023).

Wang, M.-H. (2017). Artificial intelligence and subfields. *Santa Clara University, Department of Computer Engineering*. https://www.cse.scu.edu/~m1wang/ai/AI_subfields.pdf.

Watson, G. (1987). Free action and free will. *Mind*, *96*(382), 145–172.

Case Study 1
Story of a Human Unicorn

The coworking space of 2023 is a blend of functionality and adaptability, designed to cater to a diverse cohort of professionals ranging from freelancers to small enterprise teams. The layout is flexible, with open desks, private offices, and meeting rooms, allowing for both collaborative and private work environments. Technology is seamlessly integrated with high-speed internet, state-of-the-art conferencing facilities, and smart booking systems for workspaces and meeting rooms. A focus on sustainability is evident with energy-efficient lighting, recycling systems, and green indoor plants contributing to a healthy and inviting work atmosphere. Amenities like cafes, wellness centers, and event spaces offer a balanced work-life environment, fostering community engagement and personal well-being alongside professional growth.

But how about a virtual coworking space? Are they possible? Let's discuss what one can look like. The notion of coworking and collaboration evolves beyond the physical space into a meticulously crafted virtual environment. This digital domain, accessible globally, embodies collaborative innovation where individuals and teams converge to foster a culture of shared knowledge, creativity, and productivity.

Upon entry, one is greeted by an intuitive, user-centric digital avatar, tailored to individual preferences and work requirements. The layout, devoid of geographical constraints, nurtures seamless human interaction, making collaboration feel as natural as a face-to-face discussion. The hallmark of this virtual collaborative space is the symbiotic relationship between humans and artificial intelligence. AI serves as a tool and also as a collaborator and a mentor, facilitating complex problem-solving, providing data-driven insights, and aiding in real-time ideation. The interaction with AI is designed to be intuitive, becoming an extension of the collaborative process rather than a hindrance. A core ethos of this environment is continuous learning and skill enhancement. Integrated within are learning systems propelled by AI, guiding individuals towards evolving their skills and staying abreast of industry advancements. This personalized learning path is intertwined with the individual's work, ensuring a harmonious blend of work and learning. In this envisioned virtual coworking space, the line between work, learning, and personal growth is seamlessly blended, and the pursuit of professional excellence and personal growth is a communal endeavor.

CAMPUS AI

But does a place like this exist? Campus AI (https://www.campus.ai) is a pioneering initiative that epitomizes the modern-day collaborative virtual innovation district. This initiative, meticulously crafted by Aureliusz Gorski, brings together universities, companies, creators, entrepreneurs, mentors, and learners into a singular, cohesive ecosystem.

DOI: 10.1201/9781032656618-7

The identity of Campus AI is a social learning platform and a virtual space that is designed to empower individuals to refine and apply their skills autonomously.

The allure of Campus AI is both in its state-of-the-art technological infrastructure and in its holistic, multifaceted approach to AI education and application. Unlike traditional learning platforms, Campus AI uniquely combines artificial intelligence into every facet of its offerings, transforming it into a learning ecosystem. Here, AI is not only a subject to be studied; it becomes a tool, a co-creator, and a facilitator.

THE BUILDING ANALOGY

Visualize Campus AI as a multistory modern building, a towering hub of innovation and learning. Each floor represents a different dimension of AI applications. You start at the ground floor where workshops and events happen. One floor up brings you to the Prompt Crafting School. This foundational course in prompt engineering serves as your initiation into the world of AI. With over 12,000 usable prompts that are continually expanding through a community-driven approach, this course lays the groundwork for your AI journey. Climbing up is an AI Gym where you can get advanced level AI courses and learn new skills. Further up is AI Makerspace where equipped with new knowledge, you will be able to build new AI-based products and applications. Finally, AI District Club is a space where AI-augmented humans congregate to brainstorm ideas and ventures.

COLLABORATION MEETS INNOVATION

The platform encourages cross-disciplinary collaborations, allowing you to work on various projects that integrate AI into fields as diverse as journalism, the arts, programming, and data science. The range of courses and projects goes beyond technical skills, venturing into the areas of AI-generated art, music, and even wellness programs. Here, you can learn to create AI-generated podcasts, compose music for the AI Radio, contribute to an AI Magazine, or create an AI-based product of your choice. The platform allows for a continuous learning loop. After completing a course on AI-generated podcasts, for example, you can collaborate with other members to actually produce a series of episodes for the AI Radio. If you are more visually inclined, you could take a course on AI and graphic design, then contribute to the AI Magazine with your own designs or articles. The projects you engage in not only enhance your portfolio but also contribute to a broader community of learners and creators.

THE TOKEN ECONOMY

Campus AI includes a token-based reward system that adds another layer of dynamism to the platform. As you engage with the platform, whether by completing courses, contributing to collaborative projects, or generating new course content, you earn tokens. These tokens serve dual functions that are critical to the ecosystem. First, these tokens act as digital badges of accomplishment. Verified through blockchain technology, they provide an immutable record of your achievements, competencies, and contributions. This is especially important in an educational landscape that is increasingly pivoting away from traditional credentials and towards skill-based qualifications. You are no longer just a passive learner accumulating knowledge; you are now an active participant earning a stake in your educational journey. Second, these tokens function as a virtual currency within Campus AI. They can be used to access premium content, enroll in specialized courses, and exchange for services within the Campus AI marketplace. In essence, the tokens you earn can be invested back into your learning or leveraged in ways that provide tangible value.

By implementing this token-based reward system, Campus AI does more than just gamify education; it fundamentally transforms it into an entrepreneurial endeavor. Each token earned is a point on a scoreboard and a stake in a larger intellectual and financial ecosystem. This incentivizes individual achievement and collective advancement. Everyone is encouraged to contribute, and those contributions are materially valued.

MONETIZE AS YOU LEARN

What sets Campus AI apart is its game-changing model of economic equity. In today's digital landscape, one's intellectual contributions on social media platforms often go uncompensated, effectively making content creators unpaid laborers in the information economy. Campus AI disrupts this status quo by instituting a revenue-sharing mechanism that rewards creative output.

Here's how it works: when your AI-assisted content gains traction, whether it is a groundbreaking research paper, a viral podcast episode, or a compelling art piece, Campus AI does not just pat you on the back; it puts money in your pocket. Specifically, 25% of the revenue generated by your content is channeled back to you. This is not just a gesture of goodwill; it is a concrete financial incentive that serves to foster a culture of innovation and creativity. The remaining 75% of the revenue is reinvested into the platform. This ensures not only the sustainability of Campus AI but also its ongoing evolution. It allows for constant updates, new course offerings, and technological upgrades, thereby creating a virtuous cycle: as the platform grows, so do the opportunities for its community members to generate meaningful and monetizable content.

In this way, Campus AI becomes more than a learning platform; it becomes an ecosystem of empowered individuals who are not just consumers of content but also contributors to the platform's intellectual and financial capital. It becomes a model that democratizes not only education; it democratizes opportunity. And in doing so, it offers a glimpse into the future of work, education, and creative endeavor – a future where everyone has a stake and everyone reaps the rewards.

THE WIN-WIN SCENARIO

The major takeaway is that Campus AI offers a dynamic and evolving learning environment that enhances your understanding of AI and offers avenues for practical application and financial reward. In essence, it is a collection of AI-based courses created to learn and upskill. But at the same time, it is a living, breathing AI community that encourages and rewards you to learn and create. It is a vision of what education could be, fueled by the limitless possibilities of artificial intelligence. The underlying ethos of Campus AI is inclusivity. It is open to everyone regardless of age, skill level, language proficiency, or physical ability. The platform's design ensures that with the aid of technology, all users have equal access to learn, create, and contribute. This inclusive design not only broadens the spectrum of ideas and innovations on the platform but also exemplifies a business model that leverages diversity for greater creativity and productivity for all ages and abilities.

The story of Campus AI demonstrates a visionary approach to collaborative learning and financial empowerment. Through AI, Aureliusz Gorski is creating a platform and is laying the foundation for a self-sustaining ecosystem that encourages continuous learning, creation, and financial independence. The model of Campus AI reflects a forward-thinking approach to education and work, breaking down traditional barriers to entry and fostering a culture of shared growth and innovation.

HOW DID IT START?

Since 2006, Aureliusz has been involved in venture building – creating new companies within organizations, showcasing a blend of forward-thinking, technical prowess, and business acumen. His expertise spans strategy formulation, design thinking, programming, fundraising, and management, empowering him to independently prototype and assemble interdisciplinary teams. This skill set has facilitated the creation of several successful market entities, raising over $70 million in development funds and recruiting 250+ skilled specialists.

Aureliusz's career took a decisive turn in 2013, driven by the global startup movement, leading him to dedicate his energies towards fostering the innovation ecosystem in Poland. He is a co-founder of Poland's inaugural coworking network and startup accelerator (AIP Business Link), an innovation hub for scale-ups (CIC Warsaw), and a foundation promoting inclusive innovation (Venture Cafe Warsaw). His community involvement extends to serving on the Metropolitan Council, aiding in the formulation of Integrated Territorial Investment Strategies and contributing as an expert to the Warsaw2030 strategy. Recognition for his contributions came in 2022 when *Manager* magazine honored him with the Manager Award.

Harnessing over a decade of ecosystem-building experience in Poland and collaborating with global innovation district leaders, Aureliusz devised a model for crafting generative micro-ecosystems. This insight propelled him to create open technological infrastructure, exemplified in his current leadership of the educational platform, CampusAI, and the cooperation platform, Generative District. These initiatives are currently set for pilot launches in several Polish cities and aim for a global rollout next year to expedite the growth of local innovation micro-ecosystems.

The advent of generative AI underpins Aureliusz's vision, enabling to reduce digital economy entry barriers and make it inclusive for all. Aureliusz envisions harnessing a cadre of engaged individuals, tools, and educational programs to trigger an innovation district effect in every city, positioning it as a linchpin for socioeconomic advancement (Figure C1.1).

FIGURE C1.1   Aureliusz Górski, the founder of CampusAI.

In 2015, Aureliusz brought the Cambridge Innovation Center (CIC) to Poland. CIC is a pioneering coworking space and innovation hub that originally started in Cambridge, Massachusetts, near the Massachusetts Institute of Technology (MIT). Since its inception in 1999, CIC has become synonymous with entrepreneurial dynamism and cutting-edge innovation. It serves as a nexus for startups, growth-stage companies, investors, and other stakeholders in the innovation ecosystem, providing a fertile environment for collaboration, creativity, and growth. Beyond physical infrastructure, CIC cultivates a vibrant community of entrepreneurs, mentors, investors, and professionals from diverse industries. It hosts a variety of events, workshops, and programs aimed at fostering collaboration, skill-building, and networking. The ultimate goal is to accelerate innovation and facilitate the journey from idea to impact.

In early 2020 as the COVID-19 pandemic hit, Aureliusz started musing about a virtual collaborative space, which would have all the elements of a prominent urban collaborative center but extend its borders across all geographies virtually. Three years later, in early 2023, this idea came to life due to the advances in generative AI technologies.

RECIPE FOR SUCCESS

In October 2023, just two weeks before the much-anticipated launch of Campus AI, we had the privilege of meeting with Aureliusz Górski at the CIC in Cambridge, Massachusetts. The setting could not have been more fitting. CIC, a global hub for innovation and entrepreneurship, provided the perfect backdrop for our discussion with Aureliusz, who himself is a luminary in the innovation landscape.

Aureliusz has earned his reputation as a "one-person unicorn," a term rarely used but absolutely fitting in his case. Not only did he conceptualize Campus AI, but he also brought it to life entirely on his own, an extraordinary feat in today's complex tech environment. He emphasized that creating Campus AI was not only about leveraging advanced technology but about cultivating an ecosystem where artificial intelligence could be democratized for learning, creativity, and business.

It started with Midjourney. When he initially sought to create a front-image design for the virtual Campus AI building, he reached out to a design company for a quote. The company came back with a $40,000 estimate and a six-month timeline. But who has time for that?

Undeterred, he decided to take matters into his own hands. Armed with an initial image idea, he read all literature available then on visual prompting. Then, after experimenting with about 300 prompts, he got the result. Within just two days, he had designed the front image for Campus AI himself (Figure C1.2).

In early 2023, MidJourney was not particularly adept at rendering logos. Aureliusz took his design into Photoshop, adding the final touches and the Campus AI logo. The result was a striking image that not only served as the virtual building's front but also encapsulated the spirit of innovation and accessibility that Campus AI stood for (Figure C1.3).

A professional design company was so impressed with what Aureliusz could design using generative AI and decided to join the project. They took up the challenge to materialize his vision into a 3D format, and a stunning, three-dimensional rendition of the Campus AI virtual building was delivered in just three days. This

FIGURE C1.2   The initial CampusAI design created in Midjourney.

FIGURE C1.3   Campus AI design with final touches created in Photoshop.

rapid execution clearly beat the $40,000 and six-month solution proposed earlier by another firm (Figure C1.4).

The swift completion of the 3D design in just five days served as a catalyst for Aureliusz, igniting a new level of ambition. It confirmed his belief that, with the power of generative AI, the boundaries of what one individual could accomplish were significantly expanded. This revelation led Aureliusz to take on an audacious experiment: could he build an entire project single-handedly, augmented only by artificial intelligence?

To meet this challenge, Aureliusz turned to ChatGPT, specifically the GPT-4 model. He went a step further by customizing the chatbot to specialize in startup-related knowledge. To accomplish this, he uploaded transcripts of some of the most insightful and relevant podcasts on startups, essentially creating an "AI startup expert." By doing so, he created a highly focused resource that had been educated to understand the nuances, challenges, and opportunities within the startup landscape.

Armed with this AI expert, Aureliusz began to formulate a comprehensive plan for Campus AI. The duo – human and machine – collaborated to generate business strategies, operational plans, and potential pitfalls and their solutions. They worked as a symbiotic unit, each contributing a unique set of capabilities: Aureliusz with his seasoned entrepreneurial skills and business acumen, and AI chatbot with its data-driven insights and vast, rapidly-accessible knowledge base.

This was an entirely new paradigm of entrepreneurship. Not a solo venture, not a team of humans, but a hybrid partnership that combined the best of human intuition and artificial intelligence. The result was a proof of concept that shattered traditional norms, showcasing how the future of business could be reshaped by the strategic leverage of AI. In collaboration with his AI expert, Aureliusz meticulously crafted

FIGURE C1.4   Final 3D design of Campus AI.

a working plan that involved assembling a team of specialized experts. These roles, invaluable to the startup's success, were articulated by the AI expert and included:

1. CTO (Chief Technology Officer) – responsible for shaping the technological strategy and overseeing the implementation of AI solutions.
2. UX/UI Designer – accountable for designing and optimizing user interfaces and user experiences.
3. Educational Content Specialist – in charge of developing, editing, and updating course materials.
4. Marketing Specialist – tasked with promoting the startup and crafting marketing strategies.
5. Project Manager – oversees planning, organization, and project execution both internally and externally.
6. Customer Support and Technical Support Specialist – provides course participants with technical support and answers to their questions about educational materials.
7. HR Specialist – as the company grows, this role will help in recruitment and maintaining an effective organizational culture.

Aureliusz is a seasoned professional, expert in running and consulting businesses, and is no stranger to the art of creating collaborative spaces. However, the idea of forming a virtual consortium of experts was intriguing to him. Capitalizing on his own extensive experience and expertise, Aureliusz didn't just create these experts in ChatGPT; he trained them. Each member of this virtual consortium has been meticulously educated by Aureliusz himself, effectively infusing the team with his own deep knowledge in tech and business. In doing so, he was nurturing an ecosystem that is poised for innovation, scalability, and long-term success.

TIMELINE

Let's take a look at the entire project timeline from the initial sketch to launch.

January 2023: The Big Bang of Ideas. The original thought had matured into a concrete idea, captured in a single, compelling image that encapsulated the vision for Campus AI. Completed with Aureliusz's drawing expertise, Midjourney, and Photoshop.

February–March 2023: The Blueprint Takes Shape. During these two months, Aureliusz created and trained a set of experts necessary to run a successful startup. He engaged in an exhaustive series of expert conversations, laying the analytical foundation for the enterprise.

Mid-March, 2023: The Team Grows. Recognizing that even the grandest of visions require collaborative execution, Aureliusz brought on board a human CTO, effectively transitioning from a solo venture to a team-based endeavor.

April 2023: The Birth of Campus AI. The following month was marked by formalities – Campus AI was officially established. From the technical perspective, the backend development for Campus AI was complete.

May 2023: Designing the Future. In May, the team worked on concepts of interior design for Campus AI and brought in instructor avatars created with custom avatar technologies.

Summer 2023: Building the Backbone. Over the summer, the platform underwent further development, introducing features such as AI PromptBook, AI MultiBot, AI Toolbox and technologies for AI Radio, AI Magazine, AI Art Gallery, AI Books, and additional ecosystem components that store and generate prompts and offer automated solutions. The AI PromptBook is used in the initial AI Prompt Crafting class.

September 23: The Soft Launch. The soft launch on September 23 was remarkable. Within a week, 300 lifetime accesses were sold. In just two weeks, without a single advertising dollar spent, the numbers were nearing 800 subscribers. Considering the low cost of running the enterprise, Campus AI became profitable before its official launch.

October 16: Full-Scale Launch. We are eagerly waiting for the official launch of Campus AI!

In a business landscape where the gestation period for startups often stretches into months and years, Aureliusz's feat of conceptualizing, prototyping, and launching Campus AI within just nine months is an extraordinary accomplishment and a compelling demonstration of how AI can serve as a force multiplier for human skills and ambitions to empower human unicorns.

Aureliusz views AI not as a "do-it-all" tool but as a wand in the hands of a wizard. AI amplifies capabilities, but like any form of magic, its effectiveness is determined by the skill of the one who wields it. Using generative AI in design, he used his established expertise to bring his view to life. Using ChatGPT, he established an advisory board of experts, yet each one was trained by him to adapt to realities of the project that Aureliusz envisioned. Using generative AI, Aureliusz became a "solo human unicorn," not by outsourcing expertise to AI but by using it to magnify his already formidable skills.

Aureliusz's vision goes far beyond self-accomplishment. Recognizing that not everyone has the trifecta of skills he possesses, he designed Campus AI as a democratizing force in the AI landscape. He understood that while he could navigate the complexities of launching a startup in record time, a novice without his level of expertise would face a steep learning curve.

That is the appeal of Campus AI. It is not just an educational platform but an empowerment engine of the new age. The goal is not just to educate but to elevate – to bring everyone to a level where they can harness the power of AI to unleash their creativity, innovate, and subsequently enrich the economy. Through AI-generated courses, crowd-sourced content, and transparent reward ecosystem, Campus AI aims to break down the barriers to AI literacy and incentivize learning and collaboration. The long-term vision of Campus AI is transformational. As more people gain these competencies, we are looking at a new wave of entrepreneurs and innovators who can launch businesses, create new business models, and contribute to social good. In essence, Aureliusz is not just building a company but also laying the groundwork for an AI-augmented society.

Campus AI aims not just to educate but to redefine the landscape of human potential. Under Aureliusz's leadership, the platform serves as a launchpad for anyone – whether a student, a retiree, a seasoned professional, or an entrepreneurial spirit. This is a societal recalibration in the face of technological advancement. It offers an inclusive, adaptive learning environment where everyone is equipped to ascend in their career and personal development. The goal is not just about staying relevant but about progressing towards a state of what one might term a "human unicorn," a multifaceted, AI-augmented individual capable of extraordinary feats. What makes this all the more compelling is that this construct is enabled by the strategic use of AI. Campus AI breaks down the complex into the actionable, thereby providing a pathway for anyone to engage with AI in a meaningful way. Campus AI is a bridge to a new paradigm of collective intelligence or what we have termed "Converging Minds," which is the cornerstone of our discourse.

HOW WE WROTE THIS CHAPTER

In case you are interested in learning how this chapter was written. Here is a step-by-step description.

One of the authors met with Aureliusz in person at CIC Cambridge for a 3-hour interview. It was not intended to be that long, but the ideas were flowing and conversation was going. The interview was recorded on a phone in three 1-hour segments. Post interview, the recorded conversation was transcribed by an AI solution, converting audio into easy-to-digest notes with focal points of the conversation. The resulting transcript was compared with the human author's written notes. This blend of AI-driven transcription and human annotation formed a foundation for the chapter. The author then crafted a structured outline for the chapter, delineating the core themes, narratives, and key takeaways. This structure served as a scaffold, guiding the flow and the articulation of ideas while working on writing along with AI. The resulting work was checked for accuracy and factual information.

By the way, AI named this process "Synthesized Intellection," signifying the seamless melding of human interaction, AI-assisted transcription, and collaborative text generation towards crafting a coherent, insightful narrative.

Case Study 2
How We Wrote This Book

Harnessing the power of AI does not replace human creativity, and we believe that the journey of writing this book is an example of that. We can safely say that we, humans, wrote this book. Thus, the book is a human endeavor, enhanced by AI. While AI tools like ChatGPT, Bard and Claude assisted us, the vision, structure, and final content are human-made. On the other hand, the experience of writing together with AI shows that generative artificial intelligence can be a valuable collaborator in creative processes when used effectively. This chapter outlines how we combined human expertise with AI capabilities to bring this book to life and how exactly we used generative AI throughout the writing process.

ChatGPT version 4 by OpenAI took the heavy lifting assisting us in this process. We assigned four roles to our ChatGPT assistant: style amplifier, critic, field expert, and idea generator.

As an idea generator, ChatGPT was instrumental in brainstorming sessions. It generated ideas that we either adopted or modified, adding depth to our discussions and the book itself. As a style amplifier, ChatGPT helped refine our writing style. It generated drafts that we edited for clarity and coherence, ensuring the text aligned with our vision. As a critic, ChatGPT evaluated our content via a critical lens, challenging our ideas and helping us reconsider aspects that needed improvement. This iterative process enriched the book's content. Finally, as a field expert, ChatGPT's database was used to pull in facts and data relevant to our topics. It acted as a field expert, providing information that we then verified and incorporated in our writing.

Frequent conversations among ourselves were crucial for aligning our vision and fine-tuning the AI-generated content. We divided the chapters based on our individual expertise and interests, merging our styles where necessary. In the very beginning though, we found ourselves grappling with the stylistic choices offered by our ChatGPT assistant. The engine's propensity for excessive embellishments, elaborate language, and "fluffy metaphors" led us to a critical decision point. We chose to act as "style amplifiers," heavily editing the proposed style to align with our vision of clear, concise communication devoid of unnecessary verbiage. We employed ChatGPT in conjunction with Bing for research and content generation, while Bard and Claude served specialized functions. ChatGPT excelled in generating ideas and providing a foundational structure for our chapters. However, its utility was not universal; we found its capabilities less useful in areas requiring nuanced human understanding. Looking forward, we see immense potential in leveraging large language models (LLMs) for more complex tasks.

The writing process was not just quicker but, more importantly, qualitatively different. For example, factual accuracy was paramount in chapters discussing real people or complex mechanisms. These chapters had significant human input dealing

DOI: 10.1201/9781032656618-8

with fact-checking and accuracy of the results. Conversely, in chapters that permitted a more exploratory approach, such as dealing with conceptual or abstract themes, AI served as a collaborative brainstorming partner and contributed creative insights that might not have emerged through traditional research or individual contemplation.

Our "master-prompt" served as a guiding light throughout the writing process, ensuring that both humans and AI stayed aligned with the book's overarching theme. It was crafted with precision to navigate both human and AI contributors through the complexities of the writing process. It emphasized the importance of writing in a clear and concise manner, steering away from the initial heavily embellished style that ChatGPT generated. We made a conscious effort to steer clear of typical ChatGPT expressions, which often included terms like "landscapes" and "realms," woven into "intricate tapestry" delivered in long wordy sentences. As we embarked on this writing journey, we quickly noticed the tendency of ChatGPT to employ certain terms and metaphors that, while poetic, did not align with our goal of clear and concise communication. Words like "intricate," "crucial," and "tapestry" often found their way into the generated text, adding layers of complexity and vagueness that we found unnecessary for our purpose.

To mitigate this, we took a proactive approach in our editing process. We scrutinized the AI-generated content, stripping away unnecessary verbiage to ensure that the writing remained straightforward and factual. This was in line with our master prompt, which emphasized clarity and conciseness while being grounded in facts. Our aim was to produce a work that was not only engaging but also accessible and easy to understand for diverse audiences. By doing so, we were able to maintain a style that was both informative and to the point, without losing the reader in a maze of elaborate metaphors.

Additionally, in our collaborative process, we took the extra step of creating detailed personas for Professor Aleksandra Przegalinska and Professor Tamilla Triantoro. These personas encapsulated not only professional roles and subject matter expertise but also the specific teaching styles, target audiences, and overarching messages each of us wanted to deliver. By introducing these personas to our ChatGPT assistant, we went beyond the usual Q&A interactions with ChatGPT. The AI assistant's increasing "familiarity" with our personas made its responses more aligned with the tone, style, and content appropriate for the book. During this process, ChatGPT evolved into a nuanced collaborator capable of understanding the subtleties of our academic and professional voices.

This exercise served as a learning experience, highlighting the limitations of AI in understanding the requirements of academic and professional writing. It reinforced our human role as "style amplifiers," fine-tuning the machine-generated content to meet our exacting standards.

The latter part of the prompt was action-oriented and left intentionally open-ended. It outlined our identity and goals, which could be tailored according to the specific chapter or section we were working on. It also laid out an action plan, starting with the creation of a compelling title and a one-liner that would encapsulate the essence of our work. The prompt concluded with a stylistic directive, ensuring that the writing remained consistent in its tone and approach. In essence, our master prompt served as the backbone of our project. It was a set of guidelines and a strategic tool that ensured both human intuition and machine capabilities were effectively harnessed to produce a work of substance and vision.

For us, the journey in writing this book has indeed been enlightening. It has taught us that there are different ways to write, each with its unique challenges and opportunities. As we move forward, we are excited about the endless possibilities that human-AI collaboration holds for the future of writing.

Epilogue
Collaborative AI and the Next Era of Human-Machine Collaboration

As we enter the precipice of a new era, marked by unprecedented advancements in artificial intelligence, it is imperative to recognize the potential of human-AI collaboration. The journey we have undertaken in this book, *Converging Minds: The Creative Potential of Collaborative AI*, serves as an example of this evolving paradigm.

Collaborative AI is a technological and a sociocultural revolution. It challenges the traditional dichotomies between creator and tool, expert and assistant, by fostering a synergistic relationship that amplifies human creativity, decision-making, and problem-solving capabilities.

The use cases, examples, and insights presented throughout this book have demonstrated that when AI is designed to collaborate, rather than replace humans, it can lead to outcomes that are not just additive but multiplicative. It can inspire new ways of thinking, open doors to unexplored parts of knowledge, and challenge our preconceptions, thereby enriching our intellectual landscape.

By embodying the principles discussed in this book, we hope to have offered a theoretical framework and a practical demonstration of the creative potential of collaborative AI. The journey has just begun, and the best is yet to come.

As we close this chapter, let us open our minds to the limitless possibilities that await when human and machine intelligences converge. The future is not a zero-sum game; it is a collaborative endeavor that holds the promise of elevating human potential to unprecedented heights.

DOI: 10.1201/9781032656618-9

Index

Printed in the United States
by Baker & Taylor Publisher Services